T0348911

CLIMATE *INJUSTICE*

CLIMATE
*IN*JUSTICE

WHY WE NEED TO FIGHT GLOBAL INEQUALITY TO COMBAT CLIMATE CHANGE

FRIEDERIKE OTTO

TRANSLATION BY SARAH PYBUS

DAVID SUZUKI INSTITUTE

 GREYSTONE BOOKS
Vancouver/Berkeley/London

Greystone Books Ltd.
greystonebooks.com

David Suzuki Institute
davidsuzukiinstitute.org

Cataloguing data available from Library and Archives Canada
ISBN 978-1-77840-162-6 (cloth)
ISBN 978-1-77840-163-3 (epub)

Editing for English edition by James Penco
Proofreading by Jennifer Stewart
Jacket design by Javana Boothe
Text design by Nayeli Jimenez
Indexing by Stephen Ullstrom

Printed and bound in Canada on FSC® certified paper at Friesens. The FSC® label means that materials used for the product have been responsibly sourced.

Greystone Books thanks the Canada Council for the Arts, the British Columbia Arts Council, the Province of British Columbia through the Book Publishing Tax Credit, and the Government of Canada for supporting our publishing activities.

The translation of this work was supported by a grant from the Goethe-Institut.

Greystone Books gratefully acknowledges the xʷməθkʷəy̓əm (Musqueam), Sḵwx̱wú7mesh (Squamish), and səlilwətaɬ (Tsleil-Waututh) peoples on whose land our Vancouver head office is located.

To Richard.

For believing that I have more to say and for making me believe it too, without ever saying so in words.

Contents

1: Inequality in the Spotlight   1

I. Heat: How Climate Change Is Killing the Disadvantaged Across the World

2: A Continent off the Charts (Canada and the U.S.)   21

3: An African Phantom? (The Gambia)   50

II. Drought: How Colonialism and Racism Are Hiding Behind Climate Change

4: When Justice Dries Up (South Africa)   85

5: Poverty: The Root of the Crisis (Madagascar)   113

III. Fire: How Climate Litigation Is Pushing Back Against Disinformation

6: The End of the Rainforest (Brazil)   135

7: From Pawn to Game Changer (Australia)   157

IV. Flood: How Local Attitudes and Global Politics Are Saving and Destroying Livelihoods

8: Guilt and Responsibility (Germany)   181

9: A Country Drowning in Climate Damage (Pakistan)   201

10: What Now?   224

Acknowledgments   234

Notes   235

Index   256

1

INEQUALITY IN THE SPOTLIGHT

THE WORLD'S AVERAGE TEMPERATURE has increased by more than 1°C (1.8°F) since the beginning of the Industrial Revolution. At the start of 2023, global warming stood at 1.2°C (2.16°F). This might not seem like much, but for a planet—and, above all, its inhabitants—it makes a huge difference whether the average temperature measured across all land and water masses is 14°C (57.2°F) or 15.2°C (59.36°F). After all, it makes a huge difference whether your body temperature is 37°C (98.6°F) or 38.2°C (100.76°F). With warming of 1.2°C (2.16°F), the Earth is warmer today than ever before in the history of human civilization—warmer than any world humanity has ever known.

In the world we inhabit, the climate change associated with the Industrial Revolution accelerated dramatically toward the end of the twentieth century. It continues to accelerate rapidly today, in the first few decades of the twenty-first century. While we use the rather abstract concept of global average temperature to represent climate change, we feel it in rising sea levels, melting glaciers, and shifting seasons. Its impact is particularly noticeable in heat waves, droughts, and floods, all of which play a major role in this book. These extreme weather events are all changing, often increasing in intensity and frequency.

This doesn't mean that our planet wasn't showing fever-like symptoms a hundred years ago, or that climate change didn't force people to fight for their lives back then. The heat waves of the 1930s that turned the vast prairies of the U.S. into the "Dust Bowl"—claiming many lives and livelihoods in Oklahoma, Kansas, Texas, New Mexico, and Colorado—were hotter than they would have been without climate change.[1] The dramatic flooding of Huaraz, a city in the Peruvian Andes largely destroyed by a massive landslide in 1941, would have been less extreme and killed fewer people without climate change.[2]

Planetary fever is benchmarked against the Earth's preindustrial temperature, i.e., the average global temperature between roughly 1750 and 1900. Significant consequences can be seen with global warming of just 0.17°C (0.31°F); for those who (for example) lost their lives in the U.S. in 1934 due to this slight but significant increase in heat, 0.17°C (0.31°F) was the maximum acceptable rise in temperature. When global warming reached 0.21°C (0.38°F) in 1941, many people in Huaraz lost everything they had. Here too, the heat and consequent glacier melting drove countless people to the ultimate limit: you can't adapt if you're dead. For all those who perished more than sixty years later in the 2003 European heat wave, warming of a little more than 0.8°C (1.44°F) went considerably beyond the limit. In short, there have always been heat waves, but climate change is making them ever more dangerous, as I will show in the next two chapters. Limits have long ceased to be relevant for all the plants and animals that couldn't adapt quickly enough over the decades. And of course, that goes for people too.

Is that acceptable?

For a long time this question has never been explicitly posed—but given the increased and accelerated burning of fossil fuels, the implicit answer is clearly "Yes."

Only in the twenty-first century are societies debating the acceptable limit to the rise in the average global temperature. The 2015 Paris Agreement set the limit at "well below 2°C," or even at 1.5°C where possible.[3]

This 1.5°C (2.7°F) limit has since become a byword for climate change, and it shapes the way in which we talk about climate change and our future. In the media, in the political sphere, and in private, the 1.5°C target is on everyone's lips, the fever depicted as a symptom of the disease. Climate scientists are quoted describing 1.5°C not as a target, but as a limit, most suggesting that exceeding this limit will have disastrous consequences. Analogies are repeatedly invoked of the Titanic heading for the iceberg or of an asteroid plummeting toward Earth. Such comparisons may be helpful in illustrating the scale of the problem, but as metaphors they are totally inappropriate for the problem we're actually facing.

When we exceed the average global temperature rise of 1.5°C (2.7°F), most of the people who read this book will only find out via reports, some more enlightening than others. Others won't notice at all, because they lost everything in a flood when we hit 1.3°C (2.34°F) or died in a heat wave when global warming reached 1.4°C (2.52°F). If we view the 1.5°C target as a physical limit and nothing more, these deaths and damages will be completely invisible, as will the need to invest in adaptation: even if we reach but do not exceed the 1.5°C target, many people will find that Earth is no longer a comfortable place to live. The magical 1.5°C (2.7°F) is a compromise. A compromise between death, damage, and loss on the one side, and profits from burning fossil fuels on the other. It is a political target. It is a social limit, rather than a physical one.

Every tenth of a degree of global warming leads to ever greater loss and damage, but *who* feels these effects and *how* has very little to do with the weather and climate.

TEACHABLE MOMENTS

My research as a climate scientist is in attribution science. Together with my team I analyze extreme weather events and answer the questions of whether and to what extent human-induced climate change has altered their frequency, intensity, and duration. Extreme weather events raise very interesting and relevant questions beyond what role climate change actually plays in the weather today, so my work has expanded over the years and now includes much more than just physics. When I first began my research, most scientists claimed that these questions couldn't be answered. There were technical reasons for this—for a long time, researchers had no weather models capable of mapping all climate-related processes in sufficient detail. But there were other reasons that had less to do with the research itself. Let's imagine extreme flooding in Munich, Rome, or London and heavy rainfall in the slums of Durban on the South African coast. As we will explore in detail, how the people in these various places experience this extreme weather depends on the local economic and social conditions and, fundamentally, on their political situation. Researching weather—and thus, the role of climate change—in the way I do is always political, and this makes it an uncomfortable topic for many scientists. I believe it is important to show that both obstacles—the technical and the political—can be overcome; our climate models have gotten better and better, and we are coming to realize (in science too) that research cannot take place at a remove from the real world. I see impactful weather events as *teachable moments* that show, at a moment in time when people are paying attention to the weather, quite clearly how climate change is specifically affecting humanity and how it is felt and where.

The concept of teachable moments comes from the social sciences and refers to a point in time at which we can learn something particularly easily and well. Imagine a child experiencing snow for the first time: this is a good moment to learn a little about the different aggregate states of water. However, climate-related teachable moments prove very challenging if researchers aim to find out what exactly we learn from an extreme event—and, above all, who learns it. At first, I thought that extreme heat waves or floods would mainly tell me something about the changes in the atmosphere brought about by climate change. My thinking was that if I better understood the atmospheric effects, I would also learn about the weather in times of climate change. In fact, I learned a great deal more than that.

For example, I learned about the complicated relationship between extreme weather events and risk. To know exactly how big the risk of a drought is—where and for whom—we need a whole lot of information. Three main factors come into play: the natural hazard, our *exposure* to the hazard, and the *vulnerability* with which we approach it.

In 2022, the United Nations Office for Disaster Risk Reduction (UNDRR) defined natural hazards as natural phenomena that "may cause loss of life, injury or other health impacts, property damage, social and economic disruption or environmental degradation."[4] One such natural hazard appeared in West Africa in 2022. During the rainy season, which lasted from May to October, entire regions suffered from dramatic flooding. These floods were caused, in part, by above-average rainfall that, as my team and I discovered, was significantly more intense than it would have been without climate change.[5] The rainfall was a "natural hazard," but exacerbated so significantly by human-caused climate change that it was anything but natural.

To a large extent, these floods—particularly in Nigeria— were caused by the release of a dam in neighboring Cameroon, which flooded large parts of the densely populated Niger Delta. Although this delta region is barely a third of the size of the United Kingdom, it is home to more than 30 million people, almost half the U.K. population. The risk from rainfall is particularly high, both for the people and for local ecosystems and infrastructure such as buildings, bridges, roads, and water supply lines. Naturally, it's no secret that this region is uniquely *exposed* to weather and natural hazards. There's a reason why a dam was supposed to have been built in the Nigerian part of the delta to hold back the water. But this dam was never built;[6] given the poor infrastructure and high rates of poverty, people in this area are particularly *vulnerable*, affected much more adversely than those in other areas.

So how does weather become a disaster?

We can't say exactly how the effects of climate change vary by location and type of weather, but what is absolutely clear is that the more people are in harm's way and the more vulnerable they are, the greater their risk. We've learned a lot more in recent years about all aspects of risk. For example, it's now clear that climate change alters heat waves far more than other weather phenomena (see chapter 2). With every study that my team and I perform, we seek to answer the question of what these alterations actually mean for a small section of the global population. In these studies—known as "attribution studies" among experts—we analyze not just historical and current weather data, but also information on population density, socioeconomic structures, and basically everything we can find about the event itself to gain the most accurate picture of what happened and to whom. Only after all those steps do we ask whether climate change played a role. To do this, we work with various datasets that take into account a vast range of factors—land use, volcanic activity, natural weather variability, greenhouse gas levels, other pollutants, and much more. Broadly speaking, we use climate models to simulate two different worlds: one world with human-caused climate change and one without. We then use various statistical methods to calculate how likely or intense heat waves are in specific places, both with and without human-caused global warming. Take the Siberian heat wave of 2020: in Verkhoyansk in Eastern Siberia—one of Asia's "cold poles," long considered the coldest inhabited region on Earth—record temperatures were measured of 38°C (100.4°F). Attribution studies show that such heat would have been almost impossible there without human-caused climate change. The 40°C (104°F) heat that hit London in summer 2022 would not have happened without climate change either.

But it is vulnerability and exposure that determine if weather becomes a disaster. The effects of extreme events always depend on the context—who can protect themselves from the weather (and how) is always a major factor. This is why the term "natural disaster" is entirely misplaced, even though it keeps cropping up in the media and political discourse. Plus, multiple extreme events can take place at the same time, or one after another, and combine to form compound events; some may have nothing to do with the weather at all (such as the Covid-19 pandemic). All of these events weaken people, communities, and societies.

For example, as I will explain in chapter 5, one of our studies from 2021 showed that the food insecurity linked to the drought in southern Madagascar was caused mainly by poverty, a lack of social structures, and heavy dependence on rainfall, but not by human-induced climate change. Nevertheless, just as with the Nigerian floods, international reports talked only of the weather and climate. The international media barely mentioned that, in fact, the local infrastructure, which had remained unfinished for decades, played a decisive role in the disastrous drought.

How extreme events are reported—where the media focus their attention—doesn't just influence the responsive measures we think possible. It also influences who we see as responsible for implementing the next necessary steps. Describing extreme weather as a singular moment that tells us something about climate change, and nothing more, conceals the factors that have just as much (if not more) impact on the weather's effects—and provides politicians with a handy discussion framework as they try to divert attention from poor local decision-making and planning. As a teachable moment, extreme weather reveals much wider contexts.

There are two main reasons why infrastructure in both Madagascar and Nigeria is so lacking and often nonexistent: the sustained destruction of local social structures under European colonial rule and extreme inequality within the population—inequality between the genders, between rich and poor, between different ethnic groups. It is because of factors like these that climate change becomes such a life-threatening problem. As we will see later on, there are obvious dimensions of inequality that make people more vulnerable to extreme weather, such as poverty and lack of infrastructure, and much less obvious dimensions too.

When we research extreme weather events, we put societies in the spotlight and observe how the interplay of weather, climate, geography, information, communication, government structures, and socioeconomic conditions lead to disasters—and, above all, for whom. The main thing I have learned from extreme weather events is that the climate crisis is shaped largely by inequality and the still-undisputed dominance of patriarchal and colonial structures, which also prevent the serious pursuit of climate protection.[7] By contrast, physical changes such as heavier rainfall and drier soil have only an indirect effect.

In short, climate change is a symptom of this global crisis of inequality and injustice, not its cause.

Weather-related disasters are largely a matter of unfairness and injustice, not misfortune or fate. This applies at a local level—for example, when patriarchal structures insist that pregnant women living in traditional societies have to work outdoors in extreme heat because working in the fields for personal consumption is "women's work" (see chapter 3). Or when financial aid is paid to the male head of the family and never reaches those responsible for putting food on the table. But injustice is also apparent on a global scale. Let's consider climate science, a field dominated by white men (most with backgrounds in the natural sciences) who mainly conduct and lead studies focused on the physical aspects of the climate while disregarding numerous other issues. This is why far too few studies deal with the global interactions between social and physical changes in an evolving climate. It's no wonder, then, that we lack credible research findings that could inform us about the issues of loss and damage in global climate policy on a scientific basis. This makes it even more difficult to show how centuries of colonial practices by the Global North against the countries of the Global South continue to influence the way we live, think, and act.

Today, the neglect of most of the world's population means that they suffer the most from the climate crisis. Climate change can only be understood against this backdrop, and we won't be able to manage climate change unless we eliminate the historic dynamic of injustice, of domination and dependence, between the countries of the Global North and the Global South.

Before I started studying extreme weather events and their consequences, I wasn't aware just how much our world is still shaped by the idea of domination: the domination of the "West" over the rest of humanity, but also over the planet. This idea doesn't just manifest in the dramatic loss of life and livelihood in the Global South; it destroys lives in the Global North too.

Because of this fantasy of domination, many of us believe that for the world to be as it should, it must be a world that can—and will—burn fossil fuels. This is a world in which many people continue to eat as much meat as they want without consequences. We challenge this lifestyle far too rarely, and for many it remains the epitome of success. Even more rarely do we ask ourselves where this fantasy actually comes from and what has made it so successful. As a result, we continue to concentrate solely on the consequences of this lifestyle, while ignoring the social, political, and cultural causes. We measure greenhouse gases and global increases in temperature; we calculate the physical consequences of burning fossil energy sources and deforestation. Climate change becomes an asteroid—a physical threat that must be fought with technologies such as large dams, biofuels, and hydrogen-based flights of fancy; or by playing with numbers to offset our carbon footprints. As we do this, we forget that this isn't about the end of the planet or of humanity as a whole. The Earth will continue to exist. It will evolve, and so will many of its ecosystems. Perhaps one day it will no longer be dominated by mammals; it might become a world inhabited mostly by insects. Or perhaps humans will become the only surviving mammal, surrounded by a green desert in which finding food matters more than anything else. On the face of it, these changes are neither good nor bad. And yet climate change *is* bad from humanity's perspective—not because it endangers our survival as a species, but because it endangers our collective well-being. This is not about saving the climate or humanity. Quite simply, it is about saving human dignity and rights—for all of us.

CLIMATE, DIGNITY, AND RIGHTS

It's hardly news that climate change is mainly a problem because it damages people's dignity and fundamental human rights. In fact, it's the whole reason why we talk about it on an international level.

The United Nations Climate Change Conferences, known as COPs (Conferences of the Parties, attended by the signatories to the United Nations Framework Convention on Climate Change), have never been about polar bears or the downfall of the human race. They have always been about human lives and countless livelihoods—and, of course, about economic issues (which haven't always played the most important role, though that might be easy to overlook). This is demonstrated by (among other things) the debate on the 2°C (3.6°F) target. While this includes economic cost-benefit considerations, it is above all a political goal that doesn't take science into account at all: not a single scientific assessment has ever defended or recommended a specific target[8]—and with good reason, because setting such targets is ultimately an ethical issue. It can be expressed as a simple political question: How many more human lives, how many more coral reefs, how many more insects will we allow ourselves to lose to the short-term continued use of comparatively cheap fossil fuels in the Global North?

Because talk of "dangerous climate change" largely addresses the political question of who it puts in danger and when, the low-income countries in the Global South and the small island nations in particular have fought to reduce the target to 1.5°C (2.7°F)—not because the world might perish at 1.5°C (2.7°F), but because many people in the "emerging markets" (the cynical neoliberal name for the Global South) are already losing their livelihoods.

The formula is frighteningly simple: the richer we are and the more privileged our lives, the less susceptible we are to the physical consequences of global warming. To put it another way, those with the least suffer the most from the consequences of climate change. This may be for economic reasons, because the people affected can't take out insurance or live in badly insulated or poorly constructed houses. It may be for social reasons, if they can't access information and don't receive warnings, or lack health insurance and alternative income sources. This applies both to the global north–south axis and to unequal conditions in high-income countries.

There's also the fact that climate change amplifies problems enormously: just as the Covid-19 pandemic intensified social problems, climate change deepens existing inequality. Inequality destroys trust, solidarity, and social cohesion. It makes people less willing to commit themselves to the common good. Climate change intensifies inequality both within a particular society and at a global level. Already marginalized sections of the population are pushed even further to the global margins, and if someone is already living in unstable conditions, then they need to brace for even greater danger, maybe even conflict and war. To summarize, climate change does one thing above all else: it curtails fundamental rights. The right to life and freedom, the right to free movement, the right to property, to social security, to welfare, and not least the freedom of cultural life. These are all universally recognized human rights. The Paris Agreement is a human rights treaty, not a treaty on the protection of polar bears or on charity for the Global South.

When we measure the average global temperature or use models to project future changes to the climate system—for example, understanding the extent to which more water is evaporating worldwide and thus will fall again as rain—we only obtain abstract information and scenarios at first. But extreme events take place in people's actual lives, in the contexts in which they make urban planning decisions, discuss early warning systems, cultivate fields, and develop major infrastructure projects. Again and again, in different contexts, extreme events show us how changes to the weather interact with our social structures. They leave us in no doubt as to what climate change feels like here and now and in various places around the world—and who is (un)protected and how.

That is what this book is all about.

A COLONIAL-FOSSIL WORLD

Heat waves in North America and West Africa, droughts in South Africa and Madagascar, forest fires in Australia and Brazil, floods in Germany and Pakistan: these fundamentally different events hit societies that are battling very different problems, and they all demonstrate the role of climate change in different ways. But whether it's the U.S., Germany, or South Africa that's affected, it always proves true that the people who die are those with little money who can't readily obtain all the help and information they need.

And that doesn't have to be the case, no matter where they are.

In my opinion, the fact that this keeps on happening is due to one particular, and persistent, social narrative. The basic premise is that burning fossil fuels is essential to maintaining what we call prosperity, and that "freedom" isn't possible if we're imposing a speed limit.

If we compared modern society with the society of three hundred years ago, we would unquestioningly attribute many of the achievements of recent centuries—like access to clean drinking water—to the burning of fossil energies.[9] Historically, we associate coal, oil, and gas with democracy and Western values, identifying a causal link between charcoal briquettes and the welfare state: the one affects the other. But even when this is actually true, we always forget to point out that the reverse conclusion—one perishes, and the other goes with it—is as fatal as it is false.

I would like to show just how much this narrative—or *framing*—permeates all social levels and political decisions. The Global North and Global South both continue to argue that, for reasons of fairness, the countries in the Global South must initially have very high greenhouse gas emissions too, to ensure the growth of their economies. This completely ignores the fact that in the Global North (as well as elsewhere), the poor pay for the lifestyles of a small number of wealthy people, be it the worker who toils in the mines for metals or the city dwellers subjected to greater air pollution due to the use of private vehicles. Who says that what happens in the Global North is naturally better and must be imposed on the world?

Politicians, too, often assume multiple conflicting narratives when it comes to our need for fossil fuels. When the Canadian government, led by Prime Minister Justin Trudeau, declared on one Monday in June 2019 that Canada was in the midst of a national climate emergency, that didn't stop them from approving a massive pipeline expansion project the next day, Tuesday—a pipeline set to move 600,000 barrels of oil daily from northern Alberta to southern British Columbia. Trudeau tried to justify the expansion in part by claiming that the pipeline would fund green energy projects—building a pipeline to prevent future pipelines, as it were. Even if this were true, it would mean, at best, no change in oil usage any time in the near future. That is not what is needed in an emergency, when an urgent decrease in fossil fuels is necessary to eventually save lives in Canada and across the world—while the pipeline's main function is only to make a few rich people even richer. If this is as good as we can do, we need to ask ourselves whether human rights really are the normative foundation of our governments, as hoped for by the United Nations in 1948.

Similarly, the Covid-19 pandemic—just like every climate conference held to date—clearly showed just how much the colonial mindset continues to determine global politics today. Vaccines for the Global South? Not available. The Global North's promise to ensure 100 billion U.S. dollars for the Global South every year from 2020? Unfulfilled. The Global South is being left to grapple with the impact of climate change largely on its own, and climate adaptation measures remain more a dream than reality.

The exploitation of nature and people continues to determine our actions. It's difficult to find one word that encapsulates all the underlying structures, framings, and ideas. Criticisms of capitalism are easy to come by these days, and I am certainly no fan of the current neoliberal world in which there is no serious tax on wealth, success in the education system depends on who your parents are, and maximizing profit for the few is more important than quality of life for the many. But I don't want to start discussing the pros and cons of various economic systems; I'm not the right person for that. Instead, I want to identify framings and narratives that are just as effective in the neoliberal paradise of post-Brexit Britain as they are in South Africa or Pakistan. Even if I wrote a philosophical treatise, I doubt I could find a buzzword that perfectly describes these narratives. Are they neoliberal? Patriarchal? Colonial or postcolonial? Racist? Or are they "extractivist," given that we pull everything out of the ground, forests, and seas that we can, even if it means destroying the sources of all these treasures? Whichever term I choose, it will always fall short. To better understand the whole picture, we need not just a word, but this entire book.

Again and again, my interpretation of extreme weather events that develop into disasters shows that the legacy of

colonialism permeates everything. The influence of industry, which has become rich and powerful by burning fossil
fuels, shapes our global narrative of what constitutes a desirable life. As a consequence, our global society reinforces (rather
than overcomes) inequality in many aspects, resulting in the
extremely unjust effects of climate change. I will therefore call
this framing a colonial-fossil narrative. I know full well that this
term isn't perfect and doesn't cover everything. "Extractivist"
would be more accurate than "fossil," but it's a cumbersome
word, and it's important that we explicitly reference the oil, coal,
and gas industry. "Fossil" also hints that this narrative isn't just
ancient, but long since obsolete.

And while "fossil" incorporates the factors that caused the
climate crisis—climate change would have existed if Europe
hadn't conquered any colonies but if humans had still burned
fossil energy sources—this climate change would have looked
very different without the West's ongoing colonial mindset. In
essence, colonial-fossil climate change is therefore not a *climate*
crisis, a *climate* disaster, nor any other sort of dramatic compound noun involving the word "climate"; it is a crisis of *justice*.
This crisis of justice permeates human history and didn't just
start when climate change became an issue. But when combined with the effects of climate change, this crisis of justice has
achieved a new urgency and global dimension only indirectly
related to physics.

Climate change is a problem that has less to do with a col-
lapsing climate or other physical conditions than we might
think, and the consequences of this are wider-reaching than we
have been willing to admit. It clearly shows us that the main way
in which we currently research and fight climate change—as
a physics problem—falls far too short. Obviously, we need to
transform the way in which we obtain energy. Above all, however, we need to transform participation in social life and the

application of political and economic power—who makes decisions and how. It's difficult to say how we can achieve such a fundamental transformation. One of the first steps would certainly be to realize that even if we think we understand climate change and are taking it increasingly seriously—to the point that we have renamed it the "climate crisis"—this may not be the whole story.

In this book I will tell a different story, and not the one about the asteroid. And while it may not provide a concrete solution, it goes further than a physicist's analysis. It is also an attempt to approach climate change from a philosopher's perspective and highlight the multiple factors beyond physics we should be considering.

PART I

HEAT

How Climate Change Is Killing the Disadvantaged Across the World

2

A CONTINENT OFF THE CHARTS

Canada and the U.S.

"I was in fear of my life... I had to leave the area, because we could not find shelter, or cooling or remedy in Golden... I had to drive twelve hours through forest fires to drive to the coast, only to find... no services available for homeless or disabled people."[1]

"For every person who died from the heat dome, ten or more may have suffered heat stroke, dehydration, or other complications, including permanent, life-altering injuries... I saw more patients with heat illness during June's extreme heat than I have in my entire career."[2]

"Money is scarce for anything so luxurious as a proper fan or, heaven forbid, an air conditioner."

"There were several days where I don't remember what I did; I was just lying on the floor unable to get up."[3]

THE VILLAGE OF LYTTON is around 160 miles (or 260 kilometers) northeast of Vancouver in British Columbia, Canada. The area is home to elk, black bears, and lynx, and if you're lucky you might even see a bald eagle soaring through the air.[4] But Lytton's real claim to fame is its rafting. In the mid-nineteenth century, the area—where the Thompson and Fraser Rivers come together—was known as "The Forks." The Nlaka'pamux people called it ƛ'q'əmcín—or "Camchin," as it was later anglicized—meaning "crossing over." The Fraser runs through the province for more than 854 miles (1,374 kilometers), offering a panoramic, tranquil view of the shoreline at Lytton. Around thirty-four miles (fifty-five kilometers) further south, the legendary "Hells Gate" awaits, where the river narrows to only 115 feet (35 meters) wide. The Fraser River wends its way through the passage, where tourists visit from around the world to experience nature in all its glory.

And then, in the summer of 2021, the adventure became a nightmare. June 25 was the start of an extreme heat wave in northwestern North America. Several towns and cities in British Columbia and in the U.S. states of Oregon and Washington recorded temperatures of well over 40°C (104°F). Lytton made headlines for days on end as each high temperature broke the record set the day before.[5] The extraordinary heat weighed heavily on people throughout the Pacific Northwest. Hospitals were suddenly filled with patients affected by the heat, emergency rooms were inundated with people seeking help, emergency phone lines rang incessantly, and sudden deaths shocked the population. On June 29, Lytton reported a new Canadian record of 49.6°C (121.3°F). Just one day later, the extreme weather turned Canada's rafting paradise into the ultimate fiery inferno. The heat ignited devastating forest fires that spiraled out of control in the area. The weather remained dry, hot, and windy, and when the fire hit Lytton, it destroyed it so thoroughly that only a tiny part was spared: according to the member of Parliament for the area, "90% of the village [was] burned, including the centre."[6]

Heat waves are one of the deadliest natural hazards.

The Pacific Northwest heat wave, which lasted until July 7, hit a population that wasn't used to such extreme temperatures. Nobody had taken appropriate precautions, as they might have done in regions more used to such hot summers. They couldn't switch on fans or air conditioning because, quite simply, they didn't have any.[7]

A BEAST OF A HEAT WAVE

The heat wave that wiped Lytton almost entirely off the map was so unusual that my team and I took a closer look at once. My colleagues (from the U.S., Canada, the Netherlands, France, Germany, and Switzerland) and I wanted to find out how human-caused climate change had affected the maximum temperatures in this extreme heat.[8] We studied the region between the 45th and 52nd parallels of latitude and the 119th and 123rd meridians of longitude, encompassing the cities of Seattle and Portland in the U.S. and extending north of Vancouver in Canada. This vast area—where the heat raged most fiercely—is home to more than nine million people.

We were under great pressure to characterize the ongoing event as quickly as possible, while public attention was at its height, and published our findings on the last day of the extreme temperatures. From this point on, the wider public—at least in the U.S.—saw the Pacific Northwest heat wave as the embodiment of dramatic, human-caused climate change. Ten months later, American journalist David Wallace-Wells, who had just begun to write weekly climate columns for the *New York Times*, described it as a prime example of a totally different, new, and terrifying weather event.[9] Obviously it's important for the media to make people aware of how climate change is influencing today's weather, sometimes dramatically. When Wallace-Wells interviewed me in 2022 and quoted me saying that "we are starting to witness events that would not have been possible, and that we could have not really imagined to be possible, without climate change," he didn't misquote me, but he left out my broader, more important point: we don't currently imagine these events, but we can and we have to.

We may well have had great difficulty imagining that this heat wave would happen, but we could have and should have predicted it in the long term. The reason we don't imagine such events is because we *don't want* to imagine them happening. We calculate them on an abstract level—in the form of weather forecasts—but our imagination doesn't stretch far enough (often due to lack of experience) to picture what this means for people (and ecosystems, though this book is mainly about people). But that's exactly what we need to do to prepare ourselves much better for these kinds of heat waves—and to stop them from becoming disasters.

Even I wasn't totally clear on this back in June 2021. As the heat wave raged on and journalists from all over the world bombarded us with questions on the role of human-caused climate change, our primary focus was to figure out what kind of beast we were dealing with.

THE NATURAL HAZARD

Since 1990, the Intergovernmental Panel on Climate Change (IPCC) has published a report every six to eight years in which it uses the latest research findings to show the extent to which humanity is influencing the global climate. The IPCC published its sixth report in August 2021, clearly determining that both average and extreme heat have increased dramatically on all continents—and that this can be attributed to human-caused climate change.[10]

For years now, there has been no doubt that climate change has altered heat waves. We now also know that it has more dramatically altered heat waves than it has droughts or extreme rainfall. One reason for this is that heat waves are directly connected to the temperature in the atmosphere, while droughts and rainfall are also subject to slower intermediary processes that regulate the circulation of water between the oceans, atmosphere, and land.

The numbers are unequivocal: around the world, heat waves that would have been expected once every ten years (i.e., with a probability of 10 percent in any given year) were 1.2°C (2.16°F) hotter on average in 2021 than they would have been without climate change. And an event that now happens once every ten years would have been expected once every fifty years without climate change. This means that, on average, extreme heat has become five times more likely since the Industrial Revolution. As a direct consequence of climate change, heat that was previously very rare is now merely unusual worldwide; in some cases, heat waves that can be considered extreme today reach temperatures that were as good as impossible before.[11]

So far, so technical and abstract.

But most people don't live in an average country or average city; they live somewhere specific with weather characteristic of the area—cool and wet summers as in Seattle, or hot and dry summers like those in Delhi. We can't determine what constitutes a heat wave with a sweeping global average. Take the heat waves of 2022, for example. In South Asia, heat such as that experienced by India and Pakistan in March and April 2022 is thirty times more likely today than it was at the start of the Industrial Revolution, while Argentina's extreme December heat is now around sixty times more likely.[12] All three countries have to reckon with this natural hazard far more often than initially indicated by the average frequency expected for heat waves worldwide.[13]

And then there's London: people who'd spent their entire lives in the British capital had never experienced temperatures of more than 38°C (100.4°F) before 2022, and even 38°C had only been recorded once, during the 2003 heat wave. In Delhi, meanwhile, 38°C is hot, but not extreme by local standards. But heat feels particularly different in cities compared with the abstractly calculated global average. It is usually hotter, people have fewer ways to cool down, and shade and refreshing bodies of water aren't always available. Extreme heat is far more deadly in neighborhoods with unofficial settlements and slums (neither controlled nor protected by the state) and in poorly insulated prefabricated housing estates than it is in areas with lots of vegetation. To understand what climate change actually means for a city—and for a heat wave—we need to look at heat in the relevant context.

HEAT TODAY

The 2021 heat dome didn't just destroy a village. It turned something fundamental—our previous experiences and observations—on its head.

Insurance companies, city planners, and basically anyone with any interest in the weather use experiences and observations to figure out how often extreme weather is to be expected in a specific place. Take London's sewage system. Let's say I want to check whether its drains will reliably divert water away from the streets. To do so, they should be able to withstand once-in-a-century levels of rainfall.

To find out whether they can, I need to know how many millimeters of rain would have to fall in a day to reach this level.
Thankfully, I don't need to wait for that much rain to actually
fall. That would be extremely impractical and would sink any
planning before it even started. And there's one more snag: a
once-in-a-century event doesn't specifically occur once every
hundred years; it happens every hundred years *on average*. That
means it can happen two years in a row or go unseen for four
hundred years: a "once-in-a-century event" means that the
probability of such an event in any given year is 1 percent. It's
not all that frequent, but we need to assume it will happen occasionally to ensure that the Palace of Westminster doesn't end
up floating.

If we drew solely on the weather we've actually experienced,
one hundred years wouldn't be enough to measure once-in-a-
century London rainfall. We would have to live through multiple
centuries—and empty rain-measuring beakers at the same time
every single day over these hundreds of years—before we could
finally calculate the rainfall of an average century.

Luckily there's a much quicker way, and that begins at
Heathrow Airport, at one of London's weather stations. Rain
data has been collected there every day since 1960, and measuring beakers are still being used to this day. I used this data to
calculate that 0.8 inches of rain would have to fall in London in
one day before we could call it once-in-a-century rainfall. I was
able to calculate this thanks to statistical models that expand the
weather data we've measured—i.e., the weather experienced
and observed. The same process would be used to calculate
how high a dam needs to be built to protect against a once-in-
a-millennium flood. Naturally, these models become more
accurate the more actual measurement data there is. To put it
another way, the rarer the event, the harder it is to precisely

determine whether it occurs once in a millennium or, perhaps, once in five hundred years. After that, it becomes increasingly difficult to calculate the probability of recurrence.

We battled these same difficulties as we tackled the Pacific Northwest heat dome. Naturally, here too we extrapolated weather data based on observations from previous years—in this case, the past 120 years. We found that what Seattle, Portland, and Vancouver were then experiencing was literally *impossible*: in purely mathematical terms, with all difficulties taken into account, such heat should not have happened. The heat dome was an event only to be expected in an incredibly large number of years—i.e., we couldn't write a specific number of years into our prognosis. Quite simply, there's no way to say how rarely— or how frequently—it will occur. All we can say is that it is rarer than anything that can be meaningfully calculated, rarer than once in 100,000 years.

This presented us with a methodological dilemma. The heat dome had largely reduced Lytton to ash and rubble. We couldn't use the standard methods with which we usually determine probabilities of recurrence—the methods that form the starting point for every attribution study, the foundation of all infrastructure planning, and the daily bread of insurance companies. Not because there wasn't enough observation data, or there were gaps in the data, as is often the case with extreme rainfall—and, above all, with droughts. No, it was because this heat wave was such an extreme event that it was literally off the charts. In other words, it is only possible in the world in which we live today. In the weather of the past, documented by the weather observations of the last 120 years, heat like this was impossible.

And so the question of how much more likely climate change made this specific heat wave can be answered as follows: it made it any number of times more likely—but above all, it made it possible in the first place. Just based on this weather data, too many questions remain open to know what people in the region will have to prepare for in the future. There is much that remains completely unknown; for example, whether this is an extreme (very rare) event—even in today's climate—which would make this a simple case of unbelievably bad luck. Unbelievably bad luck would be the statistical equivalent of such a rare event. Or is this sort of heat part of the weather we will have to expect regularly now?

Scientists have two possible explanations at their disposal. The first is that a heat wave like this—even in the current climate with 1.2°C (2.16°F) of global warming—is an event with very low probability, even if heightened by climate change. The second is that the weather in northwestern North America has changed so fundamentally that there will be further consequences: this would mean not only that heat waves in the area would become hotter, but also that weather conditions would occur far more frequently that would enable heat waves in the first place. And their frequency wouldn't just increase at the speed at which greenhouse gas emissions are increasing, as has previously been observed in other parts of the world; it would increase much faster than that.

So which explanation is the right one?

To answer this crucial question, scientists work with climate models—the same models used for daily weather forecasts and with which they calculate the development of temperatures, air pressure, and rainfall based on the laws of physics. None of the models available to me and my team when the heat dome descended on this part of the world indicate that the second explanation (fundamental change) is correct; all models suggest that the Pacific Northwest heat dome was an extremely rare event—unbelievably bad luck, the first of the potential explanations. But when we published our study, we couldn't rule out the possibility that we were dealing with fundamental change. No matter how good it is, one study alone only ever tells part of the story. Nevertheless, other studies—there are now at least two more based on different climate models and calculation methods—came to the same conclusion we did: in a 1.2°C (2.16°F) climate, such a dramatic heat wave in this region is very rare on average, to be expected around once in a thousand years.[14] And this once-in-a-millennium event would not have been expected at all without climate change because, as I've explained, it could only occur in the world in which we live today.

NOT ALL NUMBERS ARE EQUAL

The media love numbers, although (or perhaps because) many people are wary of mathematics. Reports often start with unemployment figures, export surpluses, or the number of dead and injured people. Numbers help to classify and compare events, but this is precisely why they should be taken with a pinch of salt.

From the outset, the results of our study were part of the story shared about the extreme heat, but they were often distorted or abbreviated. In one press release, for example, physicians and lawyers from British Columbia (commendably) called on their government to examine the health impact of the extreme heat and prepare the population better for future events. Summarizing the results of our study, they stated that the heat wave had been made "150 times more likely" as a result of climate change.[15] The physicians and lawyers didn't cite the results entirely correctly, leaving out a crucial part; the study actually said that the event had been made *a lot more than* 150 times more likely. That's a big difference. If you're lying in bed, feverish, with a body temperature a lot higher than 37°C (98.6°F), that's very different than the thermometer showing 37°C.

But numbers are often abbreviated when disseminating knowledge. And that's not surprising, given that most people are used to taking bare numbers at face value. We read single numbers in economic studies, in reports by nongovernmental organizations (NGOs), and—particularly in the age of Covid—in media reports on medical findings, assuming that they are precise and definitive. But numbers mean something different in science. Researchers always deal in approximate values, if only because measuring heights, volumes, and quantities is a very complex matter, prone to fluctuations—just think about all those measuring beakers for checking the London sewage system. This is why scientific studies never provide a single number as a result, instead offering what is unfortunately known as the "uncertainty range." "Certainty range" would actually be far more apt: when we say that this heat has been made *at least* 150 times more likely, it means we can rule out the possibility that it is just 148 or 149 times more frequent. We are very certain that it occurs more than 150 times more frequently—not 150 times, and not less. And if a journalist, press officer, or publicist does decide to publish the result in its entirety, it gets removed by the editor because we think we have learned that in science, "uncertain" translates to "no idea." In fact, it's the opposite.

We know very precisely that the Pacific Northwest heat is extremely rare, even in the world of today. Accordingly, in a 1.2°C (2.16°F) climate, it is extremely unlikely that Lytton will experience this kind of heat again any time soon; although once-in-a-millennium events always appear somewhere in the world, they never take place in the same area on multiple successive occasions. This brings us back to the demands made of the British Columbian government. If it is so unlikely that the heat dome will turn up again in northwestern North America, at least for the time being, can we then assume that Vancouver, Seattle, and Portland now have plenty of time to develop adaptation strategies that will enable them to better protect their people in future heat waves?

We might come to this conclusion if we consider only those numbers that describe how climate change has influenced physical events to date. But this would be more than shortsighted.

VULNERABILITY AND EXPOSURE

The fact that the worst heat waves we are experiencing today and will experience in the near future were barely possible in a pre-industrial climate means very little for future risk management. People don't adapt to new circumstances relative to a fixed point in the distant past. Whether they realize it or not, they act itera-tively and gradually, continuously integrating new sources of knowledge into their expertise and improving their resilience to the changing climate by analyzing and assessing the efficacy of previous efforts.[16] At least, they do if they get the chance.

Gradual adaptation is difficult if you're suddenly exposed to completely unfamiliar extremes. Right now, we are living in a climate that is 1.2°C (2.16°F) warmer. But since emissions are rising, rather than dropping, and we are a long way from net zero, in just a few years we will reach global warming of 1.5°C (2.7°F) and we will very likely reach 1.6°C (2.88°F) or 1.7°C (3.06°F), maybe more. In a 1.5°C climate, the heat experienced by Lytton will no longer be a once-in-a-millennium event, but far more likely. In a 2°C (3.6°F) world, northwestern North America will have to reckon with these sorts of heat waves every five to ten years.[17] There's no time left.

Governments that prefer to make shortsighted decisions, implying that we still have time, must accept accusations of cynicism. Heat waves are possible in unexpected places, and not just since the effects of the Industrial Revolution have become clearly noticeable.

Other parts of the world have also experienced unusually hot summers in recent years, at least when compared with the historical climate: you might remember the heat in Siberia in 2020, in Western Europe in 2019, and throughout Europe, eastern Canada, and Japan in 2018. We can also add the heat waves in Southern Europe (2017), India (2015), Russia (2010), and again in Europe (2003) to the list. They all had two things in common: they would have been up to 4°C (7.2°F) colder without climate change. And they were extremely deadly.

HEAT KILLS

Heat waves lead to high excess mortality—i.e., a mortality rate exceeding the average number of deaths. The extreme temperatures in Europe in 2003 claimed more than seventy thousand lives alone.

Heat kills directly—for example, through heatstroke—but also indirectly by amplifying existing medical conditions in connection with respiratory and cardiovascular problems. If temperatures are as extreme as they were in the 2021 heat dome, anyone exposed to the heat over a longer period—even comparatively young and healthy people—starts to feel the effects. Some sections of the population are more susceptible than others. These include older people, little children, and people with existing medical conditions, as well as the socially isolated, the unhoused, those with limited financial options, and people who work outdoors.

In aging societies like the U.S., Canada, and Germany, the number of vulnerable people is rising accordingly. King County, in the U.S. state of Washington, which has its administrative center in Seattle, is noticing an influx of older people. King County's homeless population has also grown significantly in the last ten years and is now the third largest in the U.S.

Essentially, heat-related deaths primarily affect marginalized groups whose voices are rarely heard in society or represented at editors' desks, in lawmakers' offices, or in city planning departments. Then there's the fact that heat rarely kills people on the street; most die alone in hospital or in poorly insulated homes with no air conditioning. They die invisibly, rousing no public outcry or political pressure, such as that seen in the Covid-19 pandemic. Only public health statisticians see them, and that's only if their deaths are officially recorded in some way.

Usually it's difficult to say how many deaths are caused by heat waves. During our study on the Pacific Northwest heat dome, at least several hundred heat-related deaths were assumed in the region, but we were fairly certain that this was an underestimate.

To determine the number of people who die due to a heat wave, measures include taking the number of deaths to be expected on average for the time of year without a heat wave and deducting this from the total number of actual deaths. However, these numbers only allow for a rough estimate. Incorporating death certificates makes things more precise. But even these provide limited information: they are often not available until weeks or months after temperatures cool down, because most people who die in a heat wave don't die of heatstroke, but of cardiovascular, respiratory, and other conditions. In such cases, it's rare for heat to be mentioned as a contributing factor on the death certificate.

The U.S. Centers for Disease Control and Prevention (CDC) estimates that 702 Americans died of heat-related causes from 2004 to 2018.[18] Another report, based on data from 297 U.S. counties that form almost two-thirds (61.9 percent) of the U.S. population, determined that 5,608 people died of heat between 1997 and 2006.[19] If you compare these figures with data from countries with a similar demographic structure, such as the United Kingdom, then the American data seems to be grossly underestimated. In the United Kingdom, more than 2,500 people died of heat-related causes in 2020 alone, despite there being no heat wave that year and the U.K. population being equivalent to a fifth of the U.S. population. In Vancouver, 619 deaths were determined to be due to heat, based on death certificates for the period in which the heat dome was at its worst.[20] All data is presumably far removed from the reality. But regardless of how much these figures underestimate the actual number of heat victims, they clearly show that heat is by far the deadliest extreme weather event. In the U.S., for example, far fewer people die due to hurricanes; with the exception of 2005, when Hurricane Katrina took more than a thousand lives, most years the numbers are in the low double digits.[21]

Nobody needs to die from heat.

But that doesn't mean it's easy to avoid heat-related deaths.

ARE WAKE-UP CALLS ENOUGH?

The municipalities affected by the Pacific Northwest heat wave made varying levels of effort to save people's lives. Government bodies opened several cooling centers in Seattle, Portland, and Vancouver for unhoused people, an extremely vulnerable group who are largely dependent on government agencies and struggle to protect themselves. Helping hands gave out electrolytes, food, and water. But the cooling centers weren't just for the unhoused. They were also for those whose homes were getting too hot—and in the Pacific Northwest, that was a lot of people. Houses in this region without air conditioning are not like old European townhouses. They are extremely poorly insulated and become effectively uninhabitable—not just uncomfortably warm—once temperatures reach 40°C (104°F). A brief look at this specific situation shows just how dangerous heat can be:

Seattle is the least-air-conditioned metropolitan region in the United States; less than half of its residential areas can turn down the temperature at the touch of a button. Portland is a little better—almost four-fifths (79 percent) of its households have air conditioning—while Vancouver is worst at two-fifths (39 percent). Air conditioning is becoming more common in all three cities, and this trend is sure to continue after the 2021 heat dome. During the heat wave, government authorities provided information on sixteen different websites about how to keep cool, where cooling centers were located, and how many people they could each accommodate.

One question that has been raised since the heat dome, and above all in the months and (hopefully) years before the next dramatic heat wave, is whether these measures were enough. Were there enough cooling centers? Was enough done? Given how many people died, the answer is clearly "No." And in all probability, people didn't die because they couldn't get a place in a cooling center, but because they underestimated how unbearably hot their homes would become. By the time they thought about finding somewhere cool (if they even did think about it), it was probably too late. Or maybe they would have thought about it, if the cooling center had been nearby and easy to reach.

Those in the greatest danger often have limited mobility, and it isn't easy for them to access the news or lifesaving information. Those sixteen government websites would have been absolutely no use to my ninety-year-old aunt, for example. Even in the U.S. and Canada, two high-income countries, the internet isn't equally accessible to all, and those who need the most protection often have no access whatsoever. Information must therefore be delivered via radio and TV stations as well, as locally and in as many different ways as possible. Broadcasters also need to tell people what else they can do to keep their body temperature down and how to get help, particularly if things don't go as planned and the power goes out, as it often does in intense heat (including the 2021 heat dome).

In his novel *The Ministry for the Future*, published in the U.S. in 2020, American author Kim Stanley Robinson envisions the climate crisis as a dramatic heat wave.[22] Unbearable temperatures, power outages, no drinking water. People can barely move for the heat. First the elderly die, then the young. A dystopian world that isn't going anywhere good. But the novel isn't set in the U.S.; it takes place in India. And no wonder: at first glance, an Indian city pushed to the limit by the heat seems totally different to cold, supposedly progressive Seattle. Most North American—and perhaps European—readers probably wouldn't be able to relate to the characters or setting. Right?

None of the quotes at the start of this chapter were made up. They come from people in Vancouver and the surrounding area, discussing their experiences of surviving a disaster that was by no means natural, that was only made possible by anthropogenic (human-caused) climate change. Robinson's dystopian tale set in a seemingly far-off land became bitter reality for a section of the Canadian and American population in 2021. This natural hazard, which we have intensified, became a disaster for many people because they were vulnerable, poor, and had been disadvantaged for decades—and most still are today, provided they survived, because they exist in social and economic inequality. Because nobody—neither the general population nor the emergency responders—was sufficiently prepared for such an event. Because public communication services didn't manage to get information to those who needed to act.

Although warnings were issued in good time throughout the region by the U.S. National Weather Service and Environment Canada, the NGO Human Rights Watch leveled serious accusations at British Columbia. The regional governments may have declared a state of heat emergency, but they only informed the public health departments; the information didn't reach the general population. Early warning systems that don't work are a fundamental problem, as the people in Germany's Ahr Valley found out to their cost (see chapter 8). It simply isn't enough for the authorities to speak to each other and nobody else.

Human Rights Watch also stated that by the time the health authorities activated the heat action plans, the heat wave was already beginning to subside.[23] For many people, it was already too late. The health-care system was already overwhelmed by Covid-19, and hospital workers later compared the conditions in the emergency departments to a war zone. People begged for help and tried to persuade medical staff to go with them to see their relatives because they couldn't get through on the emergency lines.[24]

Some municipalities in British Columbia had partially formalized heat action plans, while others had only limited guidelines. These plans are known as "municipal heat response planning." Many recommend improving access to cooling facilities and point out that the most important immediate measure is to hand out drinking water. They also envisage long-term changes to the developed environment. But plans are worthless if there are no (or inadequate) implementation strategies: If it's not clear who decides when the plan comes into effect. If nobody knows who is supposed to inform whom. If plans are sitting on virtual or physical desks, or if they have been adopted by regional governments for later implementation. If people don't try them out and undergo training. Then they are useless. Paramedics, physicians, and nurses in northwestern North America had to perform tasks they had never practiced, and nobody had even considered arranging shorter shifts with higher staffing. Medical professionals aren't immune to the heat, but their shifts grew longer and longer, their workload heavier and heavier.

So does every city, every municipality, every country need its own wake-up call? France wasn't spared from the 2003 European heat wave, losing at least fifteen thousand people. It was a shock for France, and it didn't have to be, given that climate change and its consequences were well known by the turn of the millennium. But when the next, even hotter heat wave hit three years later, France was ready. It had cooling centers, drinking water, functioning early-warning systems, and flows of information—but above all, it had a population that remembered 2003 very well.

This last point is particularly important. The lack of specific plans in Canadian and American towns and cities is "thought to be due to low heat risk perceptions throughout the area, as well as a lack of local data for risk assessments."[25] Unlike in France, summer heat is still truly rare in these areas. Canada, particularly the region around Vancouver, is nothing like Provence in August. The heat action plans available in 2021 were largely based on abstract information, not on experience. It's difficult to prepare for dystopian heat when it's cold and wet. If your yard never gets hotter than 28°C (82.4°F), you're not going to install sun shading on your windows, let alone think about cooling centers or stock up on drinking water. We need a trans-regional, international exchange on how to deal with the heat waves of the future and ensure there's plenty of protection in areas completely unused to the heat.

JUSTICE

Given the brutality meted out to northwestern North America by the heat dome, local and national governments must face the provocative question of whether they failed. Sometimes, specific accusations are precisely what government representatives need to wake up and improve their policies. But even if nobody has acted wrongly in a legal sense, it doesn't change the fact that the consequences are mostly borne by those who are definitely not to blame for the disaster and are least able to protect themselves.

It's unbelievable how unfairly the effects of a heat wave are distributed. We aren't all in the same boat, especially when it's on fire. It's easy to see when you look beyond the numbers and see the political, social, and economic contexts. In the Canadian province of British Columbia, heat means one thing above all else: when someone is pushed to the margins, unable to participate in their own society, it puts their life at risk. The margins are where we find the people who lose the most to the heat.

And the media? Did they fail?

Anyone scanning popular Seattle newspapers and media in late June 2021 for information about the heat could pick up the *Seattle Times* on June 28—the fourth day of the heat wave—and learn that the heat was expected to peak in the coming day, that an outdoor pool and some ice cream parlors had been closed due to the heat, and that schools and many businesses had also decided to close. The article is illustrated with photos of people swimming in nearby Lake Union, of children splashing in the city park fountain, and of windows with handwritten notes explaining that businesses were closed due to heat.[26]

This is just one newspaper of many, but it's an example of how the media tend to report on heat waves. They don't talk to people like Edward McArthur, the unhoused man quoted at the start of this chapter who fled the town of Golden in southeastern British Columbia and traveled for hours through forest fires only to reach a place that had no space for him. Seldom do you read or hear that for many Seattle residents, heat means extreme personal danger, that it can kill them and their families.

Instead, the local press, and soon the global media, reported on the other consequences of the heat, such as the fires. The devastating blaze brought Lytton to tragic fame. The media quickly latched onto the forest fires, which aren't usually isolated incidents. Usually they occur in several places at the same time—a phenomenon not entirely unfamiliar now, which has increased in this region due to climate change.[27] When combined with the extreme dryness, which allowed the fires to spread rapidly, this meant that entire cities had to be evacuated urgently. Evacuations featured in the media for other reasons too: in some parts of British Columbia, the heat caused snow to melt extremely quickly, rapidly increasing the water levels of some lakes and rivers and forcing areas north of Vancouver to issue evacuation orders due to acute flood risk. The reports tended to end here: drama, usually with no specific protagonists, far away and noticeably impersonal.

The locals may be a little better informed—for example, if they are told to close their windows to stop the forest fire smoke from entering their homes. But this presents a tricky dilemma, as the fires are closely connected to the heat wave. Heat doesn't just increase the risk of forest fires significantly; the heat and fires intensify each other's effects and health risks. When it gets very hot, we need more oxygen to maintain a normal body temperature. We breathe quicker and more deeply, which has its own pitfalls: in dry heat, the air is very polluted, while in humid heat oxygen is displaced by the water molecules that have amassed. So if you want to breathe, you have to choose between outdoor air heavily polluted by the smoke and low-oxygen air in the hot, sticky interior.

SOCIAL TIPPING POINTS

If heat is so deadly, how can we prevent deaths? This is an existential question for a not-insignificant proportion of the population, even in one of the world's richest countries. Right here, right now, climate protection is a matter of life and death. No matter where you are.

Both Canada and Western Europe have the economic resources required to drastically reduce emissions and perform adaptation measures. For my colleague Kristie Ebi, professor of global health at the University of Washington, the 2021 heat dome is a classic example of what happens in an economically well-equipped region that has plenty of resources but is completely unprepared for these sorts of temperatures.[28] This may have proved to be the tipping point for the metropolitan region, the point at which enough information, experience, and risk awareness finally converge for things to change. The wake-up call that France appears to have heard in 2003 may have been one of these tipping points, or at least a tentative tipping point— apart from heat protection, France isn't exactly a role model in climate matters and is taking far too long to adapt to the climate and prevent emissions.

In many countries, "tipping point" is a buzzword that encapsulates climate change like no other, and yet could hardly be more misleading. It is usually used to indicate a physical phenomenon that (like the popular asteroid analogy) denotes the moment at which a system tips from one state to another. This is often understood as the breakdown of the climate system as we know it. Ebi, and a few other scientists, use it in a different way. She uses "tipping point" to describe a social dimension, targeting climate change's greatest danger: the injustice with which climate change hits those who benefit the least from our economic system and have very little way of fighting it. Looking at it this way, every further death caused by this fatal injustice will destabilize society more and more. If society wants to stabilize itself again, it has to act and prevent deaths. This is the social dimension of the tipping point: it asks whether, slowly but surely, wherever we are, we are starting to notice the sort of social adjustments that precede real change.

As is so often the case, the search for answers leads to the heart of our society's unwritten and often unconscious (and therefore all the more powerful) structures. Only by understanding how these structures are crafted, and the ideas that hold them together, can we understand climate change, its causes and consequences, and better protect humanity. It's not enough just to tackle the greenhouse gases in the atmosphere and believe that temperatures are the end of the story. We'll never get a handle on climate change that way. Fighting its consequences without addressing its causes is like wrapping a feverish patient in leg compresses without tackling the source of the infection. We have no choice but to look at where it all began: human economic activity and its associated narratives.

It's not news that we need to overhaul our economy and the way we coexist. And yet this knowledge doesn't lead to consistent action, despite constant conflicts between our notion of eternal growth and the finite nature of our resources. When we look at climate change, we see this all too clearly. If we also want to see the consequences of climate change, we need to look at the people it really affects. This reveals one thing above all else: if we want to prevent, rather than relocate, the suffering of huge swaths of the population, then we need to change the system. But this doesn't get talked about anywhere near enough, and I include myself and my colleagues in that.

We have developed a peculiar relationship with the causes and perpetrators of climate change. Imagine a city, a country, where a group of powerful people go around causing indirect but catastrophic harm to thousands of people. These people aren't on the run, and they don't hide; in fact, people know them pretty well and know exactly where they live. And yet nobody does anything. The neighbors warn each other occasionally, and sometimes we give the offenders money to perhaps cause a little less harm, but all in all they remain unchallenged.

I'm not exaggerating; RWE, an energy company based in Essen, Germany, is being paid to dig up a little less lignite in Germany, even though we know that burning any more of this, the worst-polluting type of coal, will ultimately cost many people their lives due to pollution and climate change.[29]

So the question isn't just whether we learn anything from extreme events, but what we learn and what we choose to ignore. This is a crucial point. We don't discuss climate change in a productive way. It doesn't help to castigate individual consumers—like shaming them for flying—or to latch on to the doomsday narrative while companies like RWE and Exxon-Mobil, which are aware of how much damage they are causing, are allowed to continue undisturbed. The Pacific Northwest heat dome has one particularly clear lesson to offer: we don't need to think in terms of the whole world ending. Parts of the world—usually those less in the public eye—are already being destroyed by the global injustice amplified by climate change, while life simply continues as normal for the more visible majority. And we don't change anything—or at least, far too little. Instead, a sort of habituation effect sets in.

Heat waves have already become normal for most of us. The planetary apocalypse fails to materialize; local disasters become the new normal. If you'd made a movie about the Covid-19 pandemic in the 2010s, it would have become a dystopian classic. Today, Covid is part of the reality of life. And we have long since learned to accept the pandemic and its repercussions.

Similarly, climate change is no asteroid. It is a human-caused reality that escalates the inequality and injustice in our society. An injustice we consider so normal that often we don't even talk about it. But we need to start talking. And we need to focus the debate on improving people's lives here and now. To talk about climate change is, then, to talk about inequality and injustice—and about the system in which we live.

LUNGS HAVE NO LOBBY

In rich countries with high social inequality like Canada and the U.S., we can see very clearly the principles on which our capitalist (or should I say colonial-fossil?) system is built. In the last 250 years or so, the less-privileged section of the global population has suffered the consequences of industrialization over and over again. In the nineteenth and early-twentieth centuries, factory workers and other income-dependent demographic groups experienced extreme poverty and lower life expectancy; today, we see the same inequality in the excess mortality statistics for heat waves. In the twenty-first century, these underprivileged demographic groups die of the consequences of climate change, just as they lost their lives to early industrialization and the respiratory diseases and other consequences of extreme local pollution in the late twentieth century. This comparison may be a tad mawkish, but it illustrates the reality of climate change in rich, highly industrialized countries, particularly where social inequality remains high—in the U.S., the U.K., and of course in emerging economies like India too. We can name many reasons why climate change was ignored for decades. But when asked who pays for it, there's only one answer: it's the same people who always pay. With their lives or livelihoods, with what little savings they have, and most of all with their health as we—as industrialized societies, as individuals who see infinite value in our own lives—collectively ignore them. They count only as consumers, and sometimes as voters.

We don't lock up the people who profit from the environmental harm they cause. They're too powerful, too chummy with all those who pull the strings and make the decisions. And they seek out their victims from the poor, the sick, the ones who have no lobby. This phenomenon isn't reserved for Western democracies; it permeates all forms of government and societies. It may appear less extreme in less unequal societies such as the Scandinavian countries, but the structures are the same.

The question that keeps on arising—of how many more deaths we're happy to live with—relates not just to efforts to stop the average global temperature from rising further, but also to adapting to the consequences of global warming. We know exactly how to design towns and cities so that they provide their residents with adequate cooling when it gets really hot: with lots of greenery, well-insulated walls, and window shutters. The heat action plans of cities and municipalities clearly show that we are not lacking in knowledge: we know that windows insulate better when double or triple glazing is used. We can follow the example set by Mediterranean countries over the centuries and build our homes so that they create shade, protecting us from the blazing sun as we walk. We can build towns and cities with accessible drinking water dispensers, fewer sealed surfaces, trees on every street, and green spaces on rooftops—towns and cities that are greener overall.[30] So why don't we?

Alongside elected politicians, the group in our society primarily tasked with asking difficult and unpopular questions are the journalists, pundits, and publishers. Political, social, and economic structures aren't the only things that manifest inequality in a society. Cultural structures are also involved, and many of them are shaped by the media. We can use our local newspapers to let people know which public buildings are accessible and that it's better to ventilate their homes at night and not in the midday heat. And we can use every channel at our disposal to explain that the temperature on tree-lined streets is more than 10°C (18°F) lower than on streets without trees, illustrating the point with relevant images.[31]

That's not to say that governments, authorities, and companies aren't just as responsible. We know that renaturalizing rivers prevents flooding. And yet we continue to build cheap, poorly insulated houses on flood plains. We talk about protecting the environment and climate. But nobody talks about protecting people.

This fundamental inequality isn't just visible through climate change. For some time now, "Mums for Lungs," an activist group in London, has been committed to reducing traffic in the city. They argue that high air pollution due to increased exhaust fumes is causing significant damage to the very young in particular, as their lungs are still growing. There's no doubting the scientific veracity of this argument: the same people who suffer particularly from the heat are the ones who die from air pollution, but their deaths (or their lives) have little relevance to society—even though, in absolute numbers, they form a huge group. As a scientific team established in summer 2022, one in six deaths globally are caused partly by air (and lead) pollution.[32] This is an incredibly high number. And we accept it, despite knowing very well that these countless deaths could be prevented with car-free inner cities and emission-free buses and trains. We wouldn't even need any new technology. And yet private cars and SUVs continue to drive through our city centers. Cars have a lobby, but our lungs don't.

The only arguments that matter—and those only when there's an election—are plans to prevent traffic jams. These mean very little to the poorer section of the population (for example, in London) because they take the bus (the subway is too expensive, and so are cars). People who can afford cars live in the least densely populated districts, but the large roads on which they tend to drive (or sit in traffic jams) run through the poorer areas; as a consequence, these areas have the poorest air quality, and the number of deaths attributable to pollution is far higher than one in six. And this doesn't just apply to London.

The measures we need to take to prevent and adapt to climate change are inextricably linked with our global economic system. I'm not referring to capitalism per se, but to its implementation—primarily to lobbying and the major influence of a few at the expense of everyone else.

We could teach children that weather can be deadly and show them the best way to respond to warnings and to keep their body temperature down. But often we lack the willingness, the measures, and the plans to make it happen. It is exceptionally difficult to establish robust institutions that could help with this. Humanity has painstakingly (and not always successfully) learned to make institutions efficient and relatively immune to individual interests. In *Sapiens*, the Israeli historian Yuval Noah Harari describes how every piece of progress (which large sections of society, but not all, are able to enjoy) has been accompanied by the creation of new institutions, some of which succeeded better than others in remaining independent of lobbyists and individual interests. This includes financial authorities and banks, as well as large companies that have the appearance of institutions, such as Google and Amazon. Institutions work, if we believe the story they tell.[33]

And, as always, the stories we like to believe the most are stories of growth. We still celebrate their success: since the end of 2022, the oil company Saudi Aramco, owned for decades by the Saudi Arabian state, has been the second-most-valuable company in the world, and even held the top spot in the first half of 2022.[34] A company whose business model is based exclusively on the extraction and sale of oil and gas. Aside from a few applications in the chemical industry, the only thing our economy can do with these fossil energy sources is burn them. We reward a company that inevitably fosters climate change and will eventually lead to even more deaths, even more lost livelihoods, and even greater damage to humanity and nature. We endorse a business model that should no longer exist—if we truly believe that human rights are the greatest underlying narrative of all.

3

AN AFRICAN PHANTOM?

The Gambia

"We can't endure this."[1]

"Talking about heat impacts means talking about preventable death and suffering."[2]

"I believe humans are as vulnerable as God wants. If God wants us to be sick from heat, then we will be. If God does not permit that, then nothing will happen."[3]

"When I'm not pregnant I don't feel the heat of the sun very much, but when I am pregnant, whenever I am out, I feel like my body is burning."

"The work during the dry season, the work is harder than in the rainy season. During the rainy season only hard work you do… is plowing, but in the dry season if you don't have water it will be difficult for you; you must have water for the beds."[4]

HEAT WAVE WANTED

Every year, forest fires, droughts, and floods cause immense damage—and incur costs—around the world. We can tell how much certain extreme events and regions cost by how insurance companies approach them. To cope with the consequences of extreme weather events, they frequently turn to reinsurance companies, which are better able to absorb particularly expensive claims. Reinsurers (such as Reinsurance Group of America and Lloyd's of London) help insurance companies come to grips with risks that exceed their usual budget, collecting information on extreme events and their effects in dedicated disaster databases to better assess these extraordinary risks.

But insurance companies aren't the only ones who develop and maintain such databases.

Scientists also gather information on technological and environmental disasters around the world, from extreme weather events to earthquakes and oil spills, and chart how these disasters impact local life and economic costs. The largest of these scientific databases—the Emergency Events Database (EM-DAT), established in 1988—is based at the University of Louvain in Belgium. The university's Centre for Research on the Epidemiology of Disasters (CRED), which works with the World Health Organization (WHO), launched this database mainly to deal with health issues in disaster situations. Today, the EM-DAT is practically a who's who of global disasters. According to their website, it contains "data on the occurrence and impacts of over 26,000 mass disasters worldwide from 1900 to the present day."[5] So if anyone wants to know about the damage caused by chemical accidents or sandstorms in a particular country, they'd better log in to the EM-DAT.

And that's exactly what I did when I wanted to know how many weather-related disasters there have been to date in The

Gambia, in West Africa. The results were astounding: since 1900, The Gambia has had nine droughts, ten floods, six storms, and one forest fire—but not a single heat wave. The same goes for The Gambia's neighbors. For the entire African continent, the EM-DAT reports a grand total of three heat waves in sub-Saharan Africa (one in South Africa in 2016, one in Sudan in 2015, and one in Nigeria in 2002) and five in the North African Maghreb nations.

Eight heat waves in the whole of Africa? Since 1900?

When I saw these numbers for the first time, I was convinced I must have made a mistake. For physical reasons, temperatures vary less throughout the year in tropical and subtropical climates, so it's no wonder that the temperature variations are much lower than in Europe. For example, in subtropical The Gambia, where the rainy season lasts from July to October, the highest daily temperatures are usually between 29°C (84.2°F) and 34°C (93.2°F) all year round.

But weather is chaotic everywhere. Deviations from the mean—i.e., extremes—happen even without climate change. For example, temperatures can exceed 45°C (113°F) in India, or even 50°C (122°F) in Saudi Arabia.

And when it comes to heat, we can't leave climate change out of the equation any longer. Heat waves are growing in frequency and duration, a trend that has been noticeably gathering pace every decade since the 1950s and still shows no sign of slowing down. It's no coincidence, of course; greenhouse gas emissions have been climbing higher and faster from decade to decade.[6] Temperature observations show that these trends in the rise of extreme heat are strongest in the Middle East, in South America, and in parts of Africa. East, North, and Southern Africa are way ahead, but there are significant trends across the whole continent—including West Africa, where we find The Gambia. So the database must be wrong.

But that's not all.

Heat waves are often closely linked with heat stress, a plague that has worsened in recent years and will become even more serious in the future, particularly in tropical regions. Heat stress is triggered when heat is very humid and sweating no longer cools us down. In the future, heat stress will not only be unpleasantly intense, but increasingly life-threatening. Working outdoors—as many people do in rural African regions—will then become very dangerous.[7] Research teams looking for the hot spots of extreme heat stress in lower-latitude countries have pinpointed spots in Africa south of the Sahara.[8] However, the highest global values measured for heat stress were found on the Arabian Peninsula and in West Africa, right where (according to the EM-DAT) there are apparently no heat waves.

Mind you, these are changes that have been noted up to the present day; this isn't a forecast. The heat and climate change of today and tomorrow are definitely not the same. Not anywhere in the world, and certainly not in Africa. It's not just that a once-in-a-millennium heat wave like the 2021 heat dome can occur much more frequently in a 1.5°C (2.7°F) climate—that is, in a climate 0.3°C (0.54°F) warmer than 2022. The more extreme (the more rare) heat waves are, the greater the role that human-caused climate change plays in their ever-increasing frequency. Even if we manage to stop global warming from exceeding 1.5°C, heat waves today and in the future will be much hotter and more dangerous than anything we have experienced historically. When combined with population growth, this means that by the end of this century, twenty to fifty times more people will be subjected to dangerous heat in African cities. Contrary to what my EM-DAT request suggests, Africa is a stronghold of heat risk.

This brings us back to the who's who of disaster databases: if heat waves—past, present, and future—are a fact of life in

Africa, well documented and studied by the scientific and meteorological communities, then why aren't they included in the largest disaster database?

No matter how long I searched, I couldn't find them.

Even in other databases, all I found were those three heat waves in sub-Saharan Africa.

The data isn't missing because of an inattentive team who made a mistake. It is missing because, quite simply, heat data (that is, high temperatures) are not systematically reported in most African countries. It must also be noted that in most countries, the weather service tends to be responsible for systematically recording weather extremes. They collect heat data and pass it on to the authority responsible for disaster protection in their country, and that is where the EM-DAT and major reinsurers get their information. But in many African countries, government bodies barely report on disasters. Almost every country has a weather service, but they tend to be poorly equipped and to concentrate on traditionally important weather events (droughts and floods), meaning that data is available for these events and ends up in disaster databases via government bodies, international organizations, or NGOs. It's sobering to realize that many African weather services and their governments don't take any notice of heat. Africa may be one of the hot spots for heat stress and a stronghold of heat risk, but it clearly does not have enough resources to collect and report its heat data. And if heat isn't reported, then heat waves won't be recognized, especially by the authorities. And this is where the problem lies.

FORECASTS AREN'T WARNINGS

It quickly becomes clear just how important it is to officially report heat data when we take a closer look at the people actually affected by heat waves and how well they are able to protect themselves. People living in sub-Saharan Africa—a huge area encompassing around 90 percent of the continent—are generally much more exposed to and much less protected from extreme weather conditions than European citizens. This isn't surprising; from the southern edge of the Sahara down to Cape Agulhas—the southern tip of Africa—a much larger proportion of the population works outdoors than in European regions. And that's to say nothing of the poverty, the countless unofficial settlements, and the sometimes-extreme density of buildings that affect everyday life and provide few opportunities for protection. Presumably, a great many people in sub-Saharan Africa (as in other parts of the world) die prematurely due to severe heat without being registered as heat deaths.

Let's compare this with Europe. Since the beginning of the twentieth century, eighty-three heat waves have been registered that have led to more than 140,000 deaths, making them by far Europe's deadliest weather-related disasters. The EM-DAT reports seventy-one registered premature (heat-related) deaths for the three heat waves listed for this part of Africa. This means that African countries habitually fail to report heat-related deaths. In many African countries, NGOs focused on disaster control have taken on the task of reporting victims to databases, but even they don't register victims of heat waves, the "silent killers," because officially speaking there are no meteorological heat waves. And if people hardly ever hear about heat-related deaths, they will be largely unaware that extreme heat can be this deadly.

In Europe, we saw what a lack of threat awareness can do when the 2003 heat wave claimed more than seventy thousand lives (see chapter 2). Every European weather service had forecasted the heat wave. The whole of Europe knew it was going to get very hot. But forecasting isn't the same as warning. Even in Europe—as with the Pacific Northwest heat dome—weather reports didn't mention that temperatures in excess of 30°C (86°F) can be deadly or explain how or where people could protect themselves. The situation is even more worrying in The Gambia, which doesn't even have daily weather reports, let alone temperature forecasts. Instead, information is generally provided every six months on whether the next rainy season is expected to be normal or overly dry or wet.

It was only after the wake-up call of 2003 that many European states, administrative districts, and cities introduced heat action plans and other measures to minimize risks in the next heat events. Other parts of the world experienced similar wake-up calls. For example, the Indian city of Ahmedabad introduced a heat action plan in 2011 following an extremely hot and deadly lead-up to the monsoon season in 2010. As soon as the India Meteorological Department (IMD) next announced a heat wave in the forecast, the Ahmedabad authorities took action. The media told people how to protect themselves; the city set up additional drinking water tanks all over the place; hospitals, care homes, and other social institutions instructed their staff on how to deal with extreme heat, explained the dangers to be expected, and practiced their response. This approach was a success. When an even more intense heat wave struck five years later, excess mortality dropped significantly.

But this sort of forward planning is only possible if heat waves are reported and registered by weather services. Only then can authorities react and scientists use the available data as a basis for examining local heat-wave behavior. To be effective, heat action plans must be tailored to the region they aim to protect and be combined with early heat warnings. In 2018, for example, a heat wave in Canada killed countless people when temperatures reached 34°C (93.2°F). This wouldn't even trigger the early warning systems in Ahmedabad because its population, and those of other Indian cities, have adapted to temperatures far higher than 30°C (86°F), both physiologically and in their social and cultural structures and habits (such as wearing loose, light-colored clothing, avoiding direct sunlight, and ensuring access to free drinking water).

And in The Gambia? Well, they don't even have an early-warning system. It would be a step in the right direction if weather forecasts were introduced, local heat waves were systematically documented, and their effects were researched. Authorities in vast swaths of Africa have never reported heat death rates—and how would they, when heat waves aren't even registered? This is why we don't even know the local temperature values that would lead to heat-related deaths. The consequences of this are fatal; The Gambia, Kenya, and all of Africa's other countries need to know the values for their region before they can protect their societies appropriately by adapting to climate change. In short, if the populations of African countries want to keep pace with climate change, *best practices* are required for the heat. And these begin by systematically acknowledging heat waves and taking them seriously.

It's not immediately clear why heat waves on the world's second-largest continent have been and are ignored by the public in Africa and worldwide, given that they cause so much suffering and so many deaths. But it's still worth taking a closer look at the factors that stop Africa's heat from being taken as seriously as it should be. In my opinion, four factors are particularly significant.

Factor One: Physics

The first factor is the physical nature of heat waves themselves; for example, the temperature at which a heat wave is declared varies from country to country. The threshold is 25°C (77°F) in Scotland, but 28°C (82.4°F) in London. If Africa were to report heat waves, the temperature threshold would often be just 3–5°C (5.4–9°F) above the average temperatures; this is already unusually hot for regions at comparatively low latitudes. Compared with heat waves in the Antarctic—which had a temperature deviation of 39°C (70°F) in 2022—heat waves in the tropics seem ridiculously slight. But this is deceptive; what counts is the deviation from the mean—not in absolute numbers, but relative to the usual fluctuations. And at tropical latitudes, heat doesn't subside after peaking for a few days—it lingers for much longer than we are used to in Europe or North America. Sometimes these slow burners are so closely connected with other weather extremes that it's very difficult to see them at all, as they disappear behind whatever you think you saw at first glance. In 1992, for example, a vast region of Southern Africa—including Mozambique, Zimbabwe, Botswana, Lesotho, and most of South Africa—recorded temperatures of more than 3°C (5.4°F) above the average for a period of four months. This was clearly a heat wave for Southern Africa, even if it wouldn't have been in Europe (in terms of deviation from the mean; in absolute temperatures, it would have roasted Europeans). And yet it wasn't recognized as such—definitely not by scientists from regions at higher latitudes where countries are used to much greater temperature deviations. It probably didn't occur to them that a deviation of "just" a little over 3°C could be a heat wave. In the public consciousness, the high temperature was subsumed by an event that did make it into the global media: one of the farthest-reaching droughts ever known in the region.

But heat waves and droughts are fundamentally different weather events. Heat can be humid or dry, while droughts are always bone-dry. In the droughts that have always formed part of everyday life in sub-Saharan Africa, the rainy season either fails completely or it rains much less than the average. And unlike with heat waves, African weather services definitely report drought data. The population, governments, and NGOs all recognize and fight to overcome the humanitarian crises that these dry periods often bring.

Studying heat specifically would prove crucial here. As research has shown in recent years, what makes heat waves so dangerous beyond the immediate impact on human health is that they can trigger or amplify a cascade of other weather and climate-related events (and other consequences), dramatically worsening the impact of the heat overall.[9] For example, increased temperatures often cause water to evaporate faster. Drinking water then becomes scarce and the ground dries out, intensifying or triggering droughts in turn—this certainly played a key role in Southern Africa in 1992. In recent years, this phenomenon has been clearly seen in connection with the serious drought in the Horn of Africa. Without the high temperatures, the drought of 2022 wouldn't have been a drought at all.[10]

High temperatures also lead to meltwater floods—the impact is particularly dramatic when glaciers melt. Africa isn't exactly made of glaciers, which can only be found in the east at Mount Kilimanjaro (Tanzania), Mount Kenya (Kenya), and the Rwenzori Mountains (between Uganda and the Democratic Republic of the Congo). But things are very different in other parts of the world. In spring and summer 2022, large parts of South Asia suffered an extreme heat wave. In Gilgit-Baltistan in northern Pakistan (part of the Kashmir region, known for its breathtakingly beautiful landscape), a dam suddenly burst at a glacier lake below the Hispar Glacier in the Karakoram Mountains. Enormous floods like these are termed glacial lake outburst floods (GLOFs). Huge quantities of water, carrying pieces of ice and rubble, thunder down the mountains, sweeping away houses and bridges—in the Hunza Valley, below the Hispar Glacier, the water destroyed a crucial connecting bridge, cutting them off from the rest of the region entirely. Fields were flooded and crops and livelihoods were wiped out in an instant. Heat finds many ways to hit the most vulnerable where it hurts most, and GLOFs are one of them. At the same time, rapid snow melting depletes water supplies (also a major problem in Africa's mountain landscapes).

Heat waves don't just increase the risk of flash floods; they also increase the risk of forest fires and wildfires, just as we saw in Lytton. If you look at the satellite images used by NASA (the U.S. National Aeronautics and Space Administration) to monitor forest fires and wildfires, your eye will immediately be drawn to Central Africa.[11] It's always burning there. Many—but certainly not all—of these fires are controlled, part of agricultural practice. But even they can get out of hand.

Given the lack of data on heat waves in Africa, it's difficult to tell whether unintended fires are caused or accelerated by heat. But if heat waves take place at the same time as dry periods (as with the drought of 1992), the danger of forest fires increases significantly. Since the weather conditions required for the two simultaneous events are becoming more prevalent thanks to our burning of fossil fuels—particularly south of the Sahara—it's unlikely that heat plays no role in these fires, even if it hides behind a drought.[12]

Controlled and uncontrolled fires drastically intensify the dangers of a heat wave, particularly due to air quality, which deteriorates rapidly during wildfires and increases the number of illnesses and deaths. Things get even worse when landfill sites (for example) start to burn alongside ecosystems. As Delhi fought a heat wave in 2022, a huge landfill site caught fire and burned for at least nine days. For almost the whole of April (twenty-nine days) the air quality recorded was "very unhealthy," earning it the dubious title of Delhi's worst month for air quality.[13] Let's look at the numbers: on a summery May day in London, a city that (as we saw in the previous chapter) isn't exactly known for its pure air, the air quality is rated 25 (good) in the city center and 64 (moderate) in Lewisham, a suburb to the southeast. In April 2022, the air values in Delhi varied between 200 and 300, classed as "very unhealthy." The only rating that can top this on the Air Quality Index (AQI) is "hazardous." During Quebec's devastating forest fires of June 2023, New York (which is hundreds of miles away) suffered two days of air quality just as poor as Delhi's April 2022 heat.

Given just how many weather- and climate-related events are heightened by heat and the consequences it brings, we can conclude that heat waves drown, dry out, burn down, suffocate, and poison. That they go unnoticed by the international public when they occur in Africa is due, in part, to their ability to hide while amplifying other immense forces. It isn't easy to get a handle on this physical factor. It won't be possible to definitively identify heat waves in Africa and protect its people better until systematic heat data research and reporting is started in multiple African countries.

Factor Two: Infrastructure

As we've already established, everything begins with the question of who is responsible for monitoring and registering the effects of extreme weather in a particular country and reporting this to the EM-DAT and similar national and international databases. This, then, is an infrastructure problem.

In most industrialized countries, national government bodies are responsible for this, reporting how many people are affected by a case of extreme weather, the estimated mortality rate, and—in some cases—even the level of economic loss. Incidentally, the prospect of economic costs showing up on a balance sheet somewhere might be why three heat waves in sub-Saharan Africa actually made it into the EM-DAT. For example, there might not have been enough cooling water for industry, or cash crops (agricultural crops planted for export) might have been destroyed. But since the EM-DAT doesn't list economic damages for any of the three heat waves, we can only speculate—particularly given that heat always causes immense economic damage that (unfortunately) is still dramatically underestimated worldwide.

Unlike in industrialized countries, information on extreme weather and its effects for many low-income countries is not reported to databases via their meteorological services, but by various NGOs that can't liaise directly with official weather services or government departments; this means there is no central body that collects and standardizes the reports to allow the information to be compared. But comparable data is precisely what researchers and government officials would need to do their work. Take air temperature readings, for example: not all air is the same, even in the same place. In a refrigerator, the air in the lower compartments is cooler; likewise, the outside temperature isn't always the same. If you walk barefoot on warm asphalt, you'll quickly notice that your feet are much hotter than your head, because the ground temperature is significantly higher than the air temperature. Official weather stations around the world have therefore agreed to measure air temperature at a standard height of 6.56 feet (2 meters).

In some low-income countries, information on heat waves and their effects is not systematically collected, standardized, or compared. What makes this so grievous is that it prevents individual regions and countries from learning from one another about how to recognize heat waves and their effects and, ultimately, how and when to warn the population. Just as bad, if not worse, is the lack of international appeals to remedy this situation. The EM-DAT team do all they can to standardize the data submitted by NGOs, but it isn't enough when the crucial step required to help those truly affected by heat waves has yet to be taken. Without sufficient, comparable data, analyses can't be performed to help identify future heat waves. If they are not identified—and this is the major issue with infrastructure—people cannot be warned, and this usually means that lots of people have to die. Most African countries (plus other nations in the Global South, and some Global North countries too) need better-equipped weather services that standardize data on heat waves and their consequences and then systematically report this to the authorities.

Factor Three: Research and Knowledge Production

The way in which we gain knowledge about heat waves also plays a part in how the population, governments, and meteorological agencies across Africa handle the heat. Global databases on extreme weather are largely managed by companies and research organizations in Western countries, where the data is also analyzed. The EM-DAT is based in Belgium, Munich Re in Munich, Statista in Hamburg and London. None of these institutions or companies have research teams outside of the West, and they tend to struggle to look beyond their own doorsteps. This is no coincidence, but a consequence of the colonial-fossil narrative that shapes our world, a fundamental reflection of the arrogance with which Western societies approach the topic of climate justice.

This has a lot to do with the actual victims of climate change.

Let's take another look at the costs of heat waves. As mentioned, we simply have no idea how high the overall costs are. Occasional studies may examine how heat affects specific aspects (such as the loss of workforce) in individual cases, but there are absolutely no systematic analyses—things would surely be different if heat waves tended to affect the supposed winners of colonial-fossil economic activity, rather than the people who lose out around the world. A good example of how things can be different is the economic cost of tropical storms in the U.S. These costs are watched closely because every year they devour huge sums from the budgets of the insurance industry. With heat waves, however, reinsurers lack any incentive to record them in their databases. The damage incurred is entered (if at all) as a consequence of other natural events, rather than connected with heat waves. The EM-DAT lists damage of around 1.5 billion U.S. dollars for the 1992 African drought, with no awareness or mention of the heat dimension hidden therein. Heat damage was presumably not incorporated comprehensively into the damage total, which largely recognizes agricultural damage and export losses.

As long as the Global South is systematically excluded from the international scientific community, unable to actively participate and ask its own questions to better understand and research heat and its consequences, the specific challenges that heat presents to the Global South will not be given sufficient attention by the research world. If we want to live in a more just and less unequal world, African countries need their own disaster databases, their own climate models, and their own research centers, and the research institutions and companies of the Global North need to start looking beyond their own horizons and responding not just in the places where they are directly affected.

Factor Four: Media

The fourth and final factor I want to highlight here, a factor that Africa has to battle as it deals with the heat, is the international media and their lack of interest in Africa as a continent of diverse and fascinating countries that share many of the Global North's problems and also have problems of their own. While extreme weather in Europe makes headlines for weeks on end—not always the right ones—reports on weather extremes in Africa are practically nonexistent. The few that are published tend to appear only briefly, illustrated with the same stock photos of agricultural workers or slum dwellers that Western journalists always use when writing about low-income countries. This "poverty porn" has nothing to do with what's actually happening and signals to Western readers that the story has absolutely nothing to do with them; either it is completely removed from their own daily realities, or it is so depressing that they will prefer to ignore it.

To take a different angle, media sensation-mongering is another sign of the West's complete lack of interest in colonial-fossil economic "losers." This may have been the reason why the Nigerian heat wave of 2002 made its way into the EM-DAT, where it is listed with a maximum temperature of 60°C (140°F). If this value were correct, Nigeria would have held a record truly worthy of headlines. However, this value has never been confirmed, as would normally be routine, suggesting that it was not recorded at the standard measurement height of 6.56 feet (2 meters), but on the ground. It may also have been a simple measurement error; mistakes and false media representations are not uncommon. During the extreme heat in India and Pakistan in 2022 (one of the heat waves at tropical and subtropical latitudes that we know a lot about), temperatures reached 50°C (122°F), but not more. At the end of April 2022, a satellite map of India made its way around Twitter showing not air temperatures, but ground temperatures. The ground temperature actually hit 62°C (143.6°F) at times, which is extremely hot, but this doesn't reflect the air temperature and distorts the picture. It's obvious that tropical African heat waves in which temperatures are 3°C (5.4°F) higher than usual (and thus, maximum temperatures are between 30°C and 40°C [86°F and 104°F]), making a crucial difference, have little chance of attracting attention over sensational temperature records.

Records are one thing. When birds fell dead from the sky during the 2022 Indian and Pakistani heat wave, Russia had just attacked Ukraine. Ukraine is one of the world's most important wheat exporters, and so global grain prices soared, hitting the "winners" of colonial-fossil economics where it hurts most. The heat wave also affected India's wheat harvest. The Indian government imposed a ban on exports to stabilize domestic prices, increasing the global price again. The Indian and Pakistani heat wave achieved tragic, and extensive, media fame because it had a significant impact on the global wheat market—forcing the Global North to look beyond its own borders. Heat waves on the African continent couldn't attract that level of fame because most African countries don't export to the global market on such a large scale.

It won't be easy to overcome all the obstacles connected to these four factors. But if we are truly to fight climate change— something I think is inevitable for our society—then our efforts must center on these factors and their causes, their underlying structural injustice and inequality. Some of the worst and most unjust aspects of capitalism manifest in climate change. Heat waves on the African continent show how this applies to the colonialist structures that remain far from rectified, and that ensure that many former colonies remain dependent on international funds and suffer under unimaginably ill-functioning administrative structures, particularly in the field of health care. For many African countries to adapt to climate change accordingly and look to a secure future, they must first overcome this administrative wasteland. It would at least be a start if we could finally stop ignoring the most dramatic changes in the weather caused by climate change—i.e., heat waves—not just for Africa and the entire Global South, but, as we will see shortly, for the Global North as well.

ADAPTATION PIONEERS AND LAGGARDS

In public discussions and climate policy debates in the Global North, the topic of adaptation is usually treated like an unpleasant relative reluctantly invited to important events—at least, it would be if it were up to automotive and energy industry lobbyists, who already feel they have to deal with "net zero" far too often. If it were up to them, adaptation would be swept discreetly under the carpet.

Former colonies, like The Gambia, see things differently. For decades, they have been advocating at an international political level for the colonial dimension of climate change to be acknowledged worldwide. They demand that political weight be afforded to the topic of climate change adaptation and to loss and damage already suffered; for example, by the Global North acknowledging its role as the main perpetrator of climate change—not just on paper, but in the form of financial support (see chapter 9). They make it absolutely clear that by continuing to ignore the reality of life in Africa, the world is putting itself in danger. Nigeria's Adenike Oladosu—an important African voice in the global fight for climate justice—alludes to the huge spaces south of the Sahara that are becoming increasingly uninhabitable, forcing people to migrate: "If Africa is not safe, definitely Europe is not safe; hence nowhere is safe until everywhere is safe."[14] And she's right. If an enormous continent like Africa is unstable, then the whole planet begins to falter.

Given that, out of all extreme weather phenomena, heat is particularly adept at killing, countries with histories as colonies that have the necessary financial and structural capacity are taking adaptation measures into their own hands. India is one example, and Pakistan, Bangladesh, and Myanmar have also successfully implemented a range of early-warning systems. In some parts they have established the basics of short-term adaptation—early heat action programs—at local and even regional levels.[15] This shows how essential adaptation measures are in saving lives; after Ahmedabad consistently introduced heat action plans following the devastating heat wave of 2010, mortality dropped by around 1,200 people per year (the heat wave cost 1,344 people their lives). Over a thousand people can now continue to live each year, people who otherwise would have died when the heat hit 40°C (104°F) or more—and unlike those who continue to die in places with comparable climates.[16]

Pioneers of adaptation like these parts of India and Pakistan are certainly not the only ones demonstrating that people don't have to die or fall ill from the heat. Most countries still refuse to get on board, but sometimes adaptation happens "accidentally"—that is, without conscious efforts such as installing air conditioning or implementing targeted heat action plans. These "collateral effects"—adaptation in passing, if you will—become apparent when entire lifestyles change, when people move from the countryside to the city and spend less time working outdoors, or when unofficial settlements give way to social housing—all of which save lives in the context of deadly heat.

Even the human body seems to learn and adjust: we are becoming a little more heat-tolerant, even if in moderation. Without this, there would be no way to explain the great disparities in heat limits between Delhi and Montreal, i.e., the point at which heat becomes dangerous to us from a purely physiological perspective. It can't be down to infrastructure alone. Various studies, in places such as the Netherlands, show that physiological adaptation has taken place in recent decades. What remains unclear, however, is how long this takes, whether we can influence our heat sensitivity specifically, and whether male and female (and intersex) physiologies adapt in different ways.[17]

So if you were hoping that self-optimization could be the answer to the heat, you're barking up the wrong tree. We're never going to walk around with our own personal shields against all the problems created by the heat. Our physiological adaptation rate is slower than climate change, particularly in recent years, and there are of course physiological boundaries. Nobody would survive a bath in boiling water. In science, a temperature of 35°C (95°F) with 100 percent humidity—roughly equivalent to a steam bath—is considered the limit of what we can tolerate. If the humidity level is lower, our bodies can withstand slightly higher temperatures.

All in all, the number of heat-related deaths has been and remains extremely high. The 1,344 deaths in Ahmedabad in 2010 are not a statistical outlier; according to the EM-DAT, around 1,300 people died from heat in Karachi, Pakistan, in 2015—and as we saw in chapter 2, this is probably a gross underestimation.

And then there's the fact that very few regions, cities, and municipalities around the world are optimally adapted to the current climate. Most are not—not in Europe, not in the U.S., and not in India, the pioneer of heat awareness. This isn't good news. As we saw in the previous chapter, 1°C (1.8°F) of global warming makes a huge difference to effective killers like heat waves; it can be the difference between a totally normal summer and a once-in-a-century event. Even a tenth of a degree of further warming can be crucial to heat. So if we're serious about adaptation, our to-do list needs to start with how we handle heat.

To see just how challenging this is, consider the following: let's say humanity decides to take an aggressive approach to heat-specific adaptation measures, and we set a goal to ensure that through 2050, no more people will die of heat than do today. To put it a little more technically, the task would be to stabilize the relative number of heat-related deaths (the heat mortality rate) at today's level until the middle of the century. If this goal seems modest, because you'd like the mortality rate to be reduced, you might be surprised to learn that just to keep the mortality rate constant, our adaptation measures would have to be so efficient that they manage to reduce the heat-related mortality rate by 75 percent by 2050 to thwart rapidly advancing climate change. In other words, to achieve a true improvement that ensures significantly fewer people die of heat in 2050 than today, even high-performance measures that reduce the current mortality rate by three-quarters wouldn't help. They need to be much more effective than that. So there's no reason to discreetly sweep the topic of adaptation under the carpet; we need to take serious action, and we need to take it now.

COLONIALISM MEETS CAPITALISM

Looking at current development projects across the world, we can clearly see how urgently action is required. The vast majority of development projects aren't up to the challenge. No dam project—no matter how well thought out, no matter if it protects whole regions from flooding—will ever save anywhere near as many human lives as heat action plans that are actually implemented, because heat claims far more lives and presents far more dangers than a flood. And even if heat triggers or amplifies droughts, drinking water dams are no help if they simply dry up (as we will see in the next chapter). In this case, people have to change their behavior long before the heat hits, with heat action plans initiated as early as necessary.

To work in a warming climate, development projects on the African continent need to include the management of extreme heat. They need to ensure that extreme heat is recognized and get early-warning systems and adaptation strategies off the ground. This is why it's so important for development ministries to absorb the scientific knowledge that heat waves *do actually happen* in Africa (and that their effects finally need to be systematically studied). And unfortunately, development aid doesn't always help the people intended. This brings us to the title of this section, "Colonialism Meets Capitalism"—that is to say, the sale of products created by German engineering that can be used to consolidate old dependencies and create new ones. This works great with dams, less so with heat action plans, which won't fill the order books of a medium-sized German company. The infrastructure aspects of heat action plans are too far in the future and don't require any major projects, just lots of smaller measures. It might sound cynical, but climate change still hasn't been efficiently incorporated into development policy, not to mention foreign policy or transport and energy policies.

Germany has played a decisive and highly regarded role in international climate and development policy through its long-term commitment in many Global South countries—for example, through the GIZ (Deutsche Gesellschaft für Internationale Zusammenarbeit, or German Agency for International Collaboration) and the KfW (Kreditanstalt für Wiederaufbau, or Credit Institute for Reconstruction). Why doesn't it use this position to direct development grants to what really matters: resilience to the very real climate risks, which are of course connected to many other development goals such as fighting hunger? Instead, current development funding focuses on infrastructure projects and on insurance (coping with damage), bringing us back to the start of this chapter. But financial aid for poor countries after a disaster doesn't solve their main problem: their extreme vulnerability.

Improved protection against climate change, becoming more resilient, is just as important in the Global South as preventing humanitarian disasters (if not more so). Equal funding should definitely be allocated for both. And when we reflect on the findings of successive scientific studies, it becomes clear just how much there is to do in both protection and prevention; adaptation measures are only successful if countries have functioning governance structures—if the people have an intact network of rules, procedures, and institutions to fall back on in which private and public players work together to achieve social goals (see chapter 5). Setting up this sort of basic framework is expensive and arduous and takes decades of financial and institutional support. Without these funds, the framework probably won't be stable enough to efficiently counter the consequences of colonialism; extreme urbanization, the growth of informal settlements, corruption, and inefficient administrative structures can't just be dismantled, and certainly not without additional funding if there's a fundamental lack of money in the country itself. The same goes for the consequences of patriarchal structures: the greater the inequality between men and women, the less climate adaptation will work. This isn't new knowledge. But it is central, and not for nothing one of the IPCC's most important conclusions on adaptation in 2022.[18]

Although my main goal in writing this book is to show that climate change isn't a scientific problem to be rectified with a technical solution, that's not to say that we don't need physics. Heat waves turn our previous understanding of extreme weather on its head and challenge our climate models, our knowledge, and our communication methods. If we better communicated what the drastic change in heat waves means and how dangerous they are (amplifying or triggering floods, droughts, and fires and becoming so much quicker and more efficiently deadly than all other extreme weather, which evolves more slowly), this could be a real game changer in the public consciousness too and get people to take climate change more seriously. The physics world—particularly but not only in Africa—needs to know a lot more about heat waves so that we can understand region-specific vulnerabilities and develop functioning local early-warning systems. Because when thinking about weather, we must never forget that the experiences of the past and adaptations already performed are insufficient reference for future protections. We couldn't explain the 2021 heat dome with the data we gleaned from previous weather; it only became possible through climate change and created a new frame of reference that will shift further as long as we continue to burn fossil fuels. Weather that seems normal today was extreme yesterday. And very soon, today's high temperatures will almost be considered cool, although they won't feel that way in the moment.

THE FEMALE FARMERS OF THE GAMBIA

Naturally, many African people know exactly how heat feels in Africa. In the village of Keneba, around two hours from the Gambian coast, the heat might be invisible (as always), but when it gets really hot everything smells different. The heat is all-pervasive, and the air smells as if it were designed not for breathing, but for cooking or baking. The women of Keneba and the neighboring villages work outdoors a lot, as they live off the food they grow themselves. They can't evade the heat, let alone escape it. If you are or are going to become a mother and you want to survive, then you have to head out into the fields and offer yourself up to the heat.

Aminata and Zainabou (both quoted at the start of this chapter) are two of the twelve women who in 2021 participated in a pilot project that took their situation seriously with literal "fieldwork." Various local and international research teams aimed to learn more precisely how the heat felt for female farmers who were either pregnant or had recently given birth as they harvested crops or performed other outdoor work.[19]

The women allowed the researchers to regularly measure their body temperature, and they explained how they deal with the heat and what they would like to happen to help them live with it better. So far unique, this project aims to better understand the dangers of heat—particularly in a region used to it—and provides exactly what we need in many African countries in particular (and also throughout much more of the world) so that heat waves can be recorded and managed in the future. With the aid of epidemiological studies, initial data is now available on the temperatures (and above all, the humidity levels) that have direct consequences for health.

But the pilot project shows far more than that. It highlights how social inequality and increasingly difficult physical conditions make life even harder for women and intensify inequality among the population. If you want to understand how sexism works, there's a lot to learn here.

All of the women who took part in the study suffered significant heat sickness symptoms due to the hard physical labor they performed outdoors every day of their pregnancies. Not only did this affect their physical health, and sometimes that of their child, it also had a considerable impact on their general wellbeing. The heat sickness also reduced their overall productivity, to the detriment of the whole family.

The effects of extreme heat on pregnant women remain fairly unclear, although these women are particularly vulnerable as a group. The few studies available on this subject all come to similar conclusions. For example, a study of pregnant agricultural workers in Florida demonstrates that dizziness and fainting are the most common symptoms of heat sickness, and many of the workers complained of severe pain, which thankfully appeared to have no direct impact on the outcome of the birth.[20] Florida isn't The Gambia, of course, but female farm workers in Florida are constrained by many of the same socioeconomic pressures. Their income is directly connected to their productivity on the farm because they have no social security or health insurance.

As a consequence, neither set of workers takes the breaks they need, even if they feel unwell, exposing themselves to extreme temperatures far beyond what they can tolerate.

In The Gambia, the gender-specific effects of climate change are particularly striking. Aminata and Zainabou take the stance that everyone is affected by climate change, but they know first-hand that people like them—people who rely on agriculture as their main source of income and who society expects to provide for their family—feel the effects the most. The Gambia earns a quarter of its gross domestic product through agriculture, and three-quarters of the Gambian population earn their money through the agricultural industry.[21] In most cases, men work fields of cash crops belonging to large corporations or agricultural cooperatives. If the harvest turns out well, they provide the household's financial foundation. A poor harvest year will have significant economic and social consequences for the family.

And this is where the gendered division of roles comes into play.

In traditional Gambian families, women are responsible for putting food on the table. This doesn't just mean that they prepare the food. If the household coffers are empty because of a poor cash crop harvest, the women must work the fields designated for personal use to ensure the whole family has enough to eat. Climate change affects them in three ways: it has a direct impact on harvests, it reduces labor in the fields (it gets so hot that they simply can't work as much), and the (already high) burden of providing food falls entirely on the women in the extreme heat—heat that is becoming ever more extreme through climate change and pushing their bodies ever closer to the limit. This is heat that they won't survive if they don't adapt adequately.

Social relationships with other women in the community can provide pregnant Gambian women with one source of support. These relationships enable them to protect themselves better and to manage particularly challenging tasks (such as watering entire fields of plants) with helping hands. Field work involves a lot of bending down, so occasional assistance can be a real relief for many pregnant workers. If they have children old enough to take on agricultural tasks, they can ask them for help too, although the children are often either at school or employed in fields where rice or peanuts are grown for export. In households where labor is strictly divided by gender—and there are many—sons are only permitted to help their fathers grow cash crops; they are not allowed to help their mothers at home. If a woman has no daughters, she will have to work the garden or rice fields alone.

Life in The Gambia is permeated by strict gender-specific rules. Women are aware of the dangers associated with the heat. They know it's important to keep their body temperature low by drinking lots of water, not working in the midday heat, and wearing loose, light-colored clothing. But that doesn't mean they can: How are they supposed to wear the most suitable clothes when the rules require them to cover their wrists and ankles with clothing so tight that it won't ride up by so much as a millimeter?

Social status and family affiliation also play a major role. Being close to your mother, sister, or fellow wife—many Gambian men live with several wives—is often an important source of support during pregnancy. Women who don't have relatives nearby are usually left to fend for themselves, as they often feel uncomfortable asking for help from anyone other than their relations.

Social status is also linked to income, and it emphasizes how poverty, sexism, and climate change reinforce one another. Women whose husbands have no work are far more reliant on field work than others. Gender roles are so strict that their husbands don't help them harvest the rice each day (rice is the most important basic foodstuff) even if they are unemployed. In periods of reduced productivity, there is no money to buy other food or to keep the family going. If the head of the household is unemployed, the family has a lower social standing, making it harder for women to ask for help. Their productivity will then drop during extreme heat, and poverty will continue to rise. Women see their decision-making opportunities dwindle further, and inequality grows within the community and between the genders.

And so while women are always the center of families and communities, they are hardly ever the focus of studies on the impact of climate change on agriculture, community, and society—despite being disproportionately affected. Wherever research and politics focus on climate change (and adaptation above all else), they ignore those who have the most to do with it. This isn't news in the social sciences. For decades, political decision-makers have ignored the fact that unyielding gender roles make people more susceptible to extreme weather conditions. Alongside the IPCC, other multilateral institutions are also highlighting the consequences of this; the United Nations Department of Economic and Social Affairs (UN DESA) has made greater gender equality one of its seventeen sustainability goals.[22]

Women have fewer resources than men, but hardly anywhere—this too is well established—are they involved in decisions about the resources they know they desperately need. If we want to lessen climate change's power *over* women, we have to ensure more power *for* women.

ADAPTATION STRATEGIES

The insights provided by the twelve Gambian women into their everyday lives emphasize the importance and necessity of thinking far beyond the physical factors of heat in adaptation measures. And not just in The Gambia—around the world, a large proportion of female workers live in similar conditions.

In the future, local research teams, hospitals, and epidemiologists can use pilot projects like this one to collaborate on precisely identifying the direct health impact of extreme heat and find out about the damage women suffer when they are subjected to ever-increasing heat as a result of climate change. Clearly, various social and economic factors shape the framework within which women make decisions. We need to understand these factors much better than we currently do. Only in this way can we get to know the values, priorities, and processes that shape people's lives locally and form the starting point for their adaptation strategies. And the fact that all the women who took part in the pilot project were highly aware of climate change speaks to the importance of women taking on a leadership role in climate change debates, particularly in The Gambia, and playing a significant part in determining political decision-making structures, not least for development aid.

So what will it take for Africa and the world (with a few pioneering exceptions) to be able to better adapt to the changing climate in the future?

In my view, this needs to be a combination of many interweaving factors. First of all, it would be important to learn how information is passed on, both locally and regionally, on the African continent—this is the only way to design efficient heat warning systems. The next step would be to find out how extreme heat affects the most vulnerable sections of the population and the extent to which this is connected to power outages and infrastructure damage, as well as what needs to happen if all machines and apparatuses were to suddenly stop working during a heat peak or if all lines were to shut down. This could be used to determine key threshold values at which heat waves should be declared in various African countries and efficient early warnings should be triggered.

It cannot be said often enough that measuring and reporting heat extremes would in itself be a huge improvement. This information wouldn't just help to reduce excessive heat mortality in general. More specifically, it would help to better understand and flag compounded climate extremes—particularly heat and drought—in one of the world's most vulnerable regions to ensure that the right warnings are issued in good time and people can be helped to take appropriate measures. The immediate introduction of early-warning systems (even if they were initially based on trigger moments developed in countries with similar climates) could improve awareness among the general population and help people learn more quickly the best course of action during local heat extremes.

FAST PACE

Since yesterday's heat is definitely not the same as tomorrow's, we don't just need to know how much heat will intensify, but also how quickly these changes will occur. In the future, we would do well to think about speeds beyond our roads. Heat definitely has no speed limit. As long as North Americans and Europeans refuse to make even laughably small lifestyle changes, the speed at which heat waves intensify won't decrease either.

Communities have different levels of resilience because they adapt to the changing climate gradually and with the aid of new knowledge sources (and, above all, experiences). High resilience is most likely to be observed in societies with accessible and widespread education; social systems that work well; lower inequality between genders, ethnicities, and generations; and government structures designed to allow civil society to participate in decisions. Societies in which these factors are less satisfactory are less resilient.

Our gradual pace of adaptation to deteriorating weather extremes—rather than changing in one great leap—can cause problems of varying degrees. It can potentially work well in public health because education and information can achieve a great deal. Not so with infrastructure. Infrastructure is a sector that moves slowly and sedately. For coastal protection measures, power grids, road and rail networks, and water supply and wastewater systems to remain effective or functioning in a warmer climate, we have to adapt them promptly to the increasing frequency and severity of heat waves.

And so the crucial question is: What do we need to do to ensure that changes to the climate don't overtake our adaptations to the climate? To find answers, we need a fairly accurate picture of the limits—be they physiological, social, or otherwise—that impair communities' ability to prepare for the unknown climate of the future. For the scientific community, this means it is no longer enough to show that climate change is influencing extreme weather conditions today. Now we need studies to find out *how quickly* climate change is progressing and what exactly that means for us.

There's no longer any point in clarifying whether a recent heat wave—say a once-in-a-decade heat wave in today's climate—would have occurred either once in a hundred or once in a thousand generations in a preindustrial climate. If the current climate has changed so much that the preindustrial world has become a poor basis for comparison, other instruments are required to understand the weather we will be exposed to in the future or the efficacy of our adaptation strategies over the past decades. After just a little over 1°C (1.8°F) of global warming, global temperature distribution has shifted so dramatically that many regions will not be able draw direct parallels with the temperature distributions of a preindustrial climate in future heat waves. Unlike with storms or floods, past weather (including the weather of recent decades) has little or no informative value for heat forecasts, as demonstrated quite emphatically by the 2021 heat dome. This is where science comes in. The weather models applicable to yesterday's world cannot be used to calculate heat wave development today, and so they must be adjusted. If we are to design functioning early-warning systems and, above all, efficient adaptation measures, our weather models must be adjusted to the new form of heat made possible by climate change.

This applies to Africa, and to everywhere else as well. We often talk about the "new normal" in climate change, meaning that we have to get used to what we are currently experiencing. But when it comes to heat, we don't know how this new normality will feel. Temperatures (and greenhouse gas emissions) have skyrocketed in the last decade alone, leaving us no time to even get close to understanding what that means. But there's one thing we know for certain: the new temperature extremes depend more strongly on our future greenhouse gas emissions than any other weather phenomenon.

PART II

DROUGHT

How Colonialism and Racism Are Hiding Behind Climate Change

4

WHEN JUSTICE DRIES UP

South Africa

"In Africa, our natural and cultural heritage defines us—it tells our story and can trace our history. Once it is lost, it can neither be replaced nor restored. How do we adapt to a lost heritage? Eventually a lost cultural and natural heritage could mean a canceled history..."[1]

"I hope Day Zero never comes, but I can see with people wanting to come here and the desperation for water, we are soon going to realize that water is more valuable than oil."[2]

"'We've had no water for two days. I'm worried about the future because it's going to be worse and worse. Because [the government] doesn't look after us,' said Elsie Hanse, 53, who lives in a shack in a township on the edge of Graaff-Reinet, a town in the Eastern Cape."

"They built us some toilets here, but we can't use them because there's no water."[3]

WHEN BRITISH JOURNALIST Jon Heggie traveled to South Africa's Eastern Cape in January 2022, what he saw would stay with him forever. As big, thick raindrops suddenly began to patter on the parched earth, a man threw himself into a puddle that had formed around a broken drain, overjoyed to see water falling from the sky. This spontaneous celebration, which later featured in Heggie's *National Geographic* article, perfectly encapsulates South Africa's relationship with rainwater.[4]

For Africa's most southerly country, rain is like liquid gold—its drinking water supply and agricultural operations are both largely dependent on rainfall. And that's a dangerous situation to be in as South Africa, which is surrounded by the Indian and Atlantic Oceans, gets drier and drier. Its precarious water supply came to global attention in 2018, when the people of Cape Town feared that "Day Zero" had come. Day Zero is the line in the sand—or in the dust—at which the city must assume its faucets will remain dry. Just one year earlier, in 2017, the situation became so fraught that the Cape Town administration expected every day to be Day Zero. Again and again, experts calculated how many days water would continue to flow if consumption remained the same and rain refused to fall, how much longer it would take until the city, home to four million people, would become the first in the world to quite literally dry up.

As I write these words, Cape Town hasn't yet reached Day Zero, but it's come pretty damn close.

And just because Day Zero has been pushed back doesn't mean it won't happen. Although private households and agricultural businesses drastically reduced their consumption, the people of Cape Town still have far less water at their disposal than they used to—a little over twenty-six gallons (around 100 liters) per person, per day. This is very little water, as we can see when we look at other countries: according to data from the

European Commission, in 2016 a U.K. citizen used forty gallons (150 liters) per day on average, and a German citizen thirty-two gallons (122 liters).[5] When the crisis intensified in Cape Town in 2018, the authorities imposed severe limitations and stopped field irrigation almost completely. For the around fifty thousand residents who relied on selling products from their irrigated gardens to survive, this wasn't just a case of withered lawns; they lost their livelihoods.

At times, citizens had just thirteen gallons (or fifty liters) of water per day for personal use. This forced them to choose between showering and washing their clothes—and to think very carefully about how many toilet flushes they should allow themselves each day (older toilet models use nine to fourteen liters per flush). Households that consumed more water than permitted were visited by officials armed with tool kits. Devices were fitted to their faucets, allowing the authorities to cut off the water supply.[6]

Cape Town's narrow escape from Day Zero in 2018—thanks to these drastic measures and plentiful rainfall as the year progressed—was a warning to the entire country. Water problems and changes in rainfall affect the whole of South Africa, not just Cape Town, as the decline in rainfall shows no signs of slowing. But the situation isn't consistent throughout the country, as its reservoirs clearly show.

These reservoirs are usually artificial lakes maintained by dams. The Western Cape, for example, has forty-four main dams with a storage capacity of more than 66 billion cubic feet (1.87 billion cubic meters). When completely full, they hold more than 475 trillion gallons (1.8 quadrillion liters) of water, equivalent to around 935,000 Olympic swimming pools.

But these reservoirs are rarely as full (close to 100 percent) as they were in August 2023, for example. The water levels of the reservoirs in the Western Cape vary greatly throughout the year and from one year to the next. Just three months earlier, in May 2023, they were only 53 percent full. The situation did improve a little at the start of June, when they were almost two-thirds full (65.5 percent), but many regions experience devastating consequences even when the reservoirs are almost three-quarters full—as they were the previous June. While Cape Town had no more restrictions than usual at this time, Oudtshoorn, a town on a long plateau in the southeast of the Western Cape, proclaimed a major emergency after local reservoirs dropped to extremely low levels. In the Eastern Cape, on the other side of the country, where Heggie watched people celebrating the rain, the reservoirs also held far too little water (just 71 percent of total capacity) despite extremely heavy rainfall at the start of the year. In other words, there was a long-term trend for most of South Africa's reservoirs to lose significant quantities of water overall, even if so far they have always filled up again in the short term. The long-term situation remains serious, as shown by the explicit request on the Western Cape Government's website: "Residents and visitors are encouraged to continue to use water responsibly."[7]

And water is as complex as it is precious: water evaporates, falls as rain or snow, trickles away into groundwater, and flows directly into rivers. If individual aspects of this cycle increase in strength—for example, if more water evaporates—all the other aspects are affected. In April 2022, the people of KwaZulu-Natal, a province located above the Eastern Cape, experienced one of the resulting extremes: rainfall was so severe in the coastal city of Durban and the surrounding area that the floods claimed more than four hundred lives and led to further dramatic damage.[8] But KwaZulu-Natal frequently battles with droughts too, the other extreme of an intensifying water cycle. As Heggie reports, rural communities south of Durban experienced their Day Zero back in 2019. This was followed by a "Day After" that lasted for weeks, a period in which they had to survive without running water and instead rely on tankers that were supposed to deliver water but didn't always show up.[9] The region's reservoirs were empty and many springs had dried up, subjecting the province's entire southern coast to dramatic water shortages as the drought continued.

As Heggie explains, in many towns and cities, including Durban, so much water is stolen from local authorities via illegal pipelines and other forms of sabotage that the supply is much worse in practice than it appears on paper—the income that towns and cities need to finance improvements to the often-poor pipeline network is literally trickling away. This is another reason why South Africa, Lesotho, and Eswatini are often just one rainy season away from Day Zero or have already been exposed to its dry reality. When the city of Gqeberha, located in the wealthy Nelson Mandela Bay municipality in the Eastern Cape, found itself confronted with the prospect of its own Day Zero in May 2022, it brought back memories of Cape Town's situation in 2018, and many South Africans asked themselves— quite rightly—whether the provincial governments had learned anything at all from the disaster.

And there's plenty to be learned—about climate change, about politics, and, above all, about justice, corruption, and racism. Cape Town may have managed to delay its Day Zero in 2018, but that doesn't mean that every person in Cape Town actually had running water. The city's drastic measures were largely conducted in poorer districts with predominantly Black populations. Many of them have been living Day Zero for a long time now, particularly as the expansion of water pipelines into these districts is constantly being pushed back.[10]

DROUGHTS IN CLIMATE CHANGE

In the last two chapters, we have seen the massive impact that climate change has on heat waves. But it doesn't stop there; climate change also affects droughts in many ways. Evaporation and rain play a central role here. Intensified evaporation—which also takes place over land, of course, and makes the ground drier—influences the distribution of rainfall; individual rain events are becoming heavier everywhere, with greater rainfall in a shorter time. This is important because torrential rain saturates the land surface more quickly, meaning that less water seeps into the deeper layers of the soil and more water runs off directly into the rivers. If the same amount of rain falls over a longer period as mild rain, the ground remains damper and groundwater reserves are maintained for longer.

Broadly speaking, changes in rainfall due to climate change can influence dry periods and droughts in three ways. First, in some areas, rainfall is becoming more intense overall; over the course of a year or season, more rain is falling than it used to. In these areas, droughts are expected to be rarer in the future, although this has so far only been observed in northern Australia and northern Europe.[11] Second, the amount of rain may remain the same overall, but it is increasingly falling as torrential rain. It can lead to more severe droughts due to the rapid ground saturation described above, longer dry periods between rainfall, and faster evaporation. Third, more infrequent (but heavier) rainfall goes hand in hand with a decrease in overall rainfall throughout the year. The places where this happens show most clearly that droughts are getting worse due to climate change, as it is here that they grow more intense and frequent. In Europe's Mediterranean region and in the Levant (the Eastern Mediterranean region—Syria, Lebanon, Israel, Palestine, and Jordan), summers have now become so dry that there's no longer anything romantic about summer rain showers. Often they lead to flash floods, such as that experienced by Emilia-Romagna in May 2023 when the hard, dried-out ground was no longer able to soak up the water. Often, the regions most affected are those already susceptible to droughts. This isn't the case worldwide, however, as droughts are just as complex as the water cycle.

There are many types of droughts and no simple answer to the question of how they are connected with climate change. Research distinguishes between and categorizes droughts based on their causes and their effects on the water cycle and water supply. First we have "meteorological droughts," when too little rain falls over a longer period. As time goes on, this leads to "hydrological droughts," when the rivers have too little water and the groundwater table drops, known as a "groundwater drought" among experts. Then we have "agricultural" and "ecological" droughts; these occur when harvests and natural ecosystems are drastically affected and flora is no longer supplied with sufficient water because the soil is too parched. These droughts, which are described in detail in the latest IPCC report, make it especially clear just how much droughts are changing due to climate change and the impact these changes are having. In addition to the Mediterranean, the risk of drought has already grown significantly in regions such as western North America, northeastern South America, Central and East Asia, southern Australia, and parts of Central, West, and Southern Africa.[12]

Agricultural and ecological droughts not only have a huge influence on nature in general, but also have a massive effect on food security—i.e., whether there is enough sufficiently uncontaminated food available for a community.[13] These droughts often hit regions that grow mainly wheat or soy, making them some of the world's most important "bread baskets." We will see what droughts mean not only for thirst, but also for world hunger, but for now I will say that wheat doesn't like extreme dryness or heat, leading to many meager harvests in recent years, as noticeably reflected for us all in increasing bread prices. If there's no wheat growing, then the (ecological) drought hits the forests, which are increasingly going up in smoke.

THREE YEARS OF DROUGHT IN CAPE TOWN

After southwestern South Africa experienced a drought so dramatic from 2015 to 2017 that by 2018 the region around Cape Town was just a few drops of water away from Day Zero, various research teams studied the drought in more detail, primarily to understand the role of climate change.[14] It was the worst drought experienced by the Western Cape since 1904 and led to unparalleled water scarcity. Historical weather observations for this region show, as I have already outlined, that there are close links between this drought and the continuous decline in rainfall. This is all the more worrying because climate models predict that long-term dryness will continue to increase throughout most of Southern Africa.

Southern South Africa is also one of the few regions in which another distinctive meteorological feature plays a role. Research anticipates that, due to increasing global warming, rainfall variability—not to be confused with rainfall distribution (from mild to torrential rain)—will decrease in this region. It won't just rain less; fluctuations in rainfall will change as well. More specifically, as various studies have shown, there won't be as many wet years with lots of rain, and years with no rain at all will occur less frequently. This doesn't seem all that dramatic at first glance, but the consequences will be fatal. This has to do with factors like municipal water management—how water is collected, protected, distributed, and secured, and how sustainably it is used.

Cape Town is a very good example of what this means in practice. Cape Town is surrounded by six large reservoirs that store fresh water for the entire region. The reason why it was so badly prepared for the three-year drought was that the city administration had opted for a water management system based on the assumption that the reservoirs would be filled in wet years. Meteorologically speaking, what happened is that in an unusually large area around the reservoirs, it rained only in winter (basically June and July) and not in spring and fall, as it usually would. A hydrological drought then occurred, which meant that the rivers and other sources were no longer feeding enough water into the reservoirs. Higher temperatures and increased evaporation presumably intensified the situation further, confronting Cape Town with a drought to be expected just once in three hundred years or so, in the climate of 2017. By the end of 2017, the water in the reservoirs was no longer sufficient to cover the needs of Cape Town's 3.7 million residents plus the irrigated agriculture in the Western Cape. This was a failure on the part of those responsible, who (as usual) had relied on wet years.

The region urgently needs to invest in resilience to future climatic extremes and make long-term plans for periods with significantly less rain. This includes reducing water consumption overall (61 gallons or 232 liters per person per day before the crisis). This can be achieved through information, but also through price incentives. But this won't be enough if the reservoirs aren't filled. Should it prove impossible to channel water to the Western Cape from various regions, making it less vulnerable to local weather fluctuations, then it might help to invest in desalination plants.

Climate change means that, in the future, Cape Town will have to reckon with more frequent dry periods of this nature, which have become around three times more likely due to global warming. In a world 2°C (3.6°F) warmer, this will become three times more likely again. Drought risks don't change exponentially here (a concept we learned as Covid-19 began to spread) but linearly; slight global warming makes droughts a little more likely, and strong global warming increases the drought risk more strongly. But don't feel relieved—this means that the probability of droughts increases continuously with every tenth of a degree of global warming.[15]

So far, so dry.

For droughts, which are notoriously difficult to study, a result like this is unusually clear. Nowhere in the world have I seen the fingerprints of climate change more clearly when it comes solely to rainfall. As we saw in the previous chapter, continuously climbing temperatures help to make droughts an ever-greater problem fueled by global warming.

The water management system of the Western Cape has a great challenge ahead of it. Alongside the Cape Town supply system, the Western Cape Water Supply System (the system for the entire Cape), with its forty-four main dams, is constructed in such a way that it is almost entirely dependent on rainfall. This makes the whole province extremely vulnerable to climatic fluctuations and changes—the water levels of the Western Cape reservoirs demonstrate very clearly how risky this is. Naturally, there are economic consequences too, both at regional and national levels: droughts place the agricultural sector under great pressure, and South Africa is already suffering from many years of fruit crop failures, losing important exports such as wine and apples. But its most important agricultural product—particularly in the Western Cape—is wheat,[16] and yields are declining there too. One of the reasons making this particularly impactful is that neighboring countries, such as extremely poor Lesotho (I'll come back to this later), rely on wheat from South Africa.

When harvests decline, it's always the poor who immediately lose their jobs and thus the financial means to cope with the effects of the drought. If you have no income, you can't buy water when the communal faucets are turned off temporarily or completely. The agricultural sector is another important official (and often unofficial) employer in the Western Cape, a region that still strongly bears the features of the apartheid regime. In South Africa's extremely unequal society, we can see just how much climate change intensifies existing inequality: in times of crisis, the first people to lose the jobs they were given under the table are those who never had a chance on the regular labor market. In the Western Cape, these are almost always people of color who have suffered institutionalized racism.

Climate change's role in the Western Cape has consequences that go far beyond the faucets of Cape Town. Above all, climate change hits a long chain of local problems, which brings us back to Cape Town's water management.

CLIMATE CHANGE IN LOCAL POLITICS

Cape Town's water crisis didn't just have physical causes such as the lack of rain. There were clearly political reasons too. Cape Town is an excellent example of the political nature of climate change.

The city manages more than just the six reservoirs previously mentioned. Together with the national Department of Water and Sanitation, the city of Cape Town manages a total of fourteen main dams and twenty-five large reservoirs, plus water pipelines. The six most important reservoirs—which supply both Cape Town (which takes around two-thirds of the water stored), the surrounding farmland, and other communities—are almost all outside of the city limits.[17] Up to now, the administration has not used any groundwater, as there is controversy surrounding whether and how it could be used sustainably for drinking water.[18] Instead, it largely draws on the local reservoirs with comparatively short pipelines. Given how Cape Town currently manages its water resources, the system is designed to function in a world in which extreme droughts happen once every fifty years on average.[19] However, a rainfall shortage that would previously have been expected every fifty years, for example, will now occur much more frequently—every seventeen years or so (that is, three times more frequently). The people and economy supplied by the Western Cape water management system are therefore much more vulnerable to droughts than previously assumed.

But in the past, the system was known for its reliability and considered a model for other South African regions. It was so reliable that very few people or businesses did anything to ensure greater independence—for example, by setting up reserve water tanks in case the municipal supply proved insufficient. No other alternative supply strategies were tested or set up either. This is unusual for cities familiar with water crises. In São Paulo, for example, industrial companies that require non-potable fresh water take water directly from water treatment plants and not from drinking water reservoirs.[20]

By making the country (which dominates Africa's southern economic area) so dependent, Cape Town—the seat of South Africa's parliament—has done South Africa an absolute disservice. Droughts in South Africa, and particularly in the Western Cape, don't just have a short-term impact on the regional and national economy; they weaken it in the long term. In their official economic report for 2017, the Western Cape authorities were forced to list the water crisis as a high financial risk for companies: rising water tariffs drastically increase costs, diminish their competitive position, and harm their reputations by calling into question their reliability and quality.[21]

And we know all too well what happens next: companies leave, unemployment rises further, existing gulfs grow wider in an extremely unequal society, social problems intensify, and trust in the local government declines further. Conversely, this means that southwestern South Africa (a rich area) needs to make its water management significantly less dependent on rainfall to protect the population. And to get companies and investors to believe in the long-term security of these plans, the region must also rebuild confidence in the economy, no easy task in a country plagued by corruption—particularly given that the city government has squandered a lot of trust since the start of the crisis, having long been a role model in water and climate management.

The true extent of the mismanagement was demonstrated by an article in the journal *Nature*, which was published while water was still at its scarcest. The author of the article, Mike Muller (Director-General of South Africa's Department of Water Affairs and Forestry until 2005), teaches at the University of the Witwatersrand in Johannesburg, where he focuses on governance and public management. In his article, entitled "Lessons from Cape Town's drought," Muller does something astonishing: he warns strongly against seeing climate change as the cause of the water crisis.[22] But Muller definitely isn't one of those climate-change deniers who only say things like this to drown a world fighting water scarcity in disinformation. Muller had another aim in mind. He wanted to show that, above all, the main problem behind the Cape Town water crisis was poor political decision-making. Was he right? I believe he was.

Muller's argument went as follows: rather than making decisions supported by science and investing to improve the water supply and safeguard it for the future, the city and water management department relied on the assumption that the measures taken in previous years to reduce overall water consumption would still be enough to get around the problem.[23]

Naturally, the next question is whether those responsible had the information they needed to make scientifically sound decisions. And yes, they did. As Muller emphasizes, South Africa's metropolitan areas have access to the highest standard of current science, allowing them to draw up the best possible water management plans and secure the system for the future. Since the 1980s, South African hydrologists (experts focused on the water cycle) have worked with international teams to develop water management models. These models are so well designed that they incorporate all key factors: rainfall quantities, rivers, reservoirs and dams, historic hydrological data, the various uses for water, and much more. In other words, the city planners who use these models consider the whole forest, not just one tree, as well as everything that grows, crawls, runs, and flies there, even who eats whom and when. These world-class models are regularly improved and enable economic and demographic developments to be predicted so that water demand can be calculated along with the storage capacity required in the reservoirs. They can map the operation of the water network in real time, simulate different risks, assess supply shortfalls for various users, and test the efficacy of improvement measures.

Cape Town has also used these sorts of elaborate models in the past. For years, the city worked closely with scientists, took a long-term view by involving other Western Cape regions, planned locations for additional reservoirs, and considered the best ways to tap into other sources—rivers that didn't currently feed the reservoirs with water. This was how the Western Cape supply system was formed, and at first it worked well.

Muller goes on to explain that in the 2000s, the construction of new reservoirs was largely stopped. This was partly due to pressure from environmentalists who launched a campaign to shift the focus to water conservation and the reduction of consumption—and to encourage more conscious regulation of water usage. This resistance delayed the completion of the Berg River Dam, the most important dam in the Cape Town catchment area, by six years. It was finally connected to the water network in 2009, and in 2018 the new reservoir played a considerable role in ensuring that Cape Town's faucets didn't run dry. While this reservoir was and remains important for Cape Town, the environmental protection campaign hit on a point also highlighted by the water management models. Rainfall is decreasing. As far back as 2009, everyone knew what problems to expect. Even the politicians.

But rather than reacting appropriately to the new situation and tapping into new sources for their reservoirs, the government spent money on other things and closed their eyes to the reality that, despite all the warnings, the six reservoirs ultimately held water reserves of less than two years' normal rain. In numbers, this is 31.4 billion cubic feet (890 million cubic meters) of water; an average annual yield is 20.1 billion cubic feet (570 million cubic meters). For a while, things appeared to go well. But appearances can be deceptive.

After the second dry winter in a row, in 2016 the city became aware of its blatantly poor planning and began to introduce the drastic measures described. Xanthea Limberg, the city councillor responsible for water services, defended the policies enacted. As Muller reports, she wrote that "it is not practical to ring-fence billions of rand for the possibility of a drought that might not come to pass."[24] But when she made this public statement, the drought that might not come to pass had been under way for two years. So it came as no surprise when this became a disaster in 2017, the third year without rain: the water management models had shown that three dry years in succession was entirely possible. The politicians and officials had all the information they needed to make scientifically sound decisions. And as Muller rightly points out, several consecutive years of drought are certainly not uncommon in the Cape Town region— it happened in the late 1930s and the early 1970s. Luckily, 2002, 2003, and 2005 (all dry years) were interrupted by a wet 2004. The hydrological models are very reliable in showing these risks, but politicians ignored them as they made their social and financial decisions in the years leading up to the 2018 water crisis. According to Muller, the water crisis was estimated to cost 181 billion U.S. dollars, including losses from declining tourism and lost jobs.

It would have been much cheaper for Cape Town to invest just one billion rand (around 70 million U.S. dollars) in infrastructure in 2013 and 2014—before the three-year drought had the region in its grip—to prepare the city for three years of potentially low rainfall. Even if it had rained in the end, any realistic financial planning team would have happily paid to insure against the possibility.

And this brings us back to the start of Mike Muller's publication. The exact thing that Muller tried to prevent with his article came true: Cape Town's decision-makers explicitly used climate change as an excuse to distract from their poor decisions. Climate change did play a role, of course, but it wasn't a surprise—it had been calculated in all forecasts, including those by the city's own hydrological models. Unlike with extreme heat (see chapters 2 and 3), historic data for droughts is very important and helpful in estimating the often-great fluctuations in rainfall from year to year in regions prone to drought. This is clearly shown by the data recorded by a weather station in the vicinity of Cape Town: between 2013 and 2017, rainfall ranged from less than 500 millimeters (19.6 inches) to up to 1,250 millimeters (49.2 inches) per year. This is similar to the difference between annual rainfall in Oslo and Madrid—that is, the difference between a wet and a Mediterranean climate.

Unlike the extreme heat in British Columbia and the northwestern U.S. in 2021, which would not have been possible without climate change, there would have been a severe drought in the Western Cape without climate change; it just wouldn't have been quite so severe. Climate change is relevant to droughts, but not a game changer as it is for heat waves, at least when it comes to rain. The fact that the people responsible appear to value populist decisions and last-minute election gifts more than a sustainable water supply for the population has nothing to do with climate change. What it shows is who their policies actually serve. In Cape Town, they clearly don't serve the people who lose their jobs first: the people who are poor, undereducated, and usually not white. Cape Town's policies are for rich white people who—as we will see—drill wells in their gardens, just in case. This is colonial-fossil thinking in a nutshell, despite the promising motto in South Africa's coat of arms: Diverse People Unite.

YET ANOTHER WAKE-UP CALL

Heat waves and droughts may differ in many ways, but persistent, severe heat and Cape Town's drought have one thing in common: it clearly takes a dramatic wake-up call for political priorities to shift. The Brazilian state of São Paulo is a striking example of how political diversionary maneuvers can intensify water problems.[25]

As in Cape Town and the Western Cape, the risk of drought is no secret in São Paulo; hydrological models and historical weather data provide all the information required. But rather than taking responsibility, in the past the city and state preferred

to concentrate on political rifts, taking years to introduce necessary measures. And then, in 2013–2014, a drought occurred
that was to be expected in São Paulo every ten years or so (rather
than once a century, for example) from a meteorological perspective. But since the country was in the throes of an election,
politicians chose not to admit what was happening. Geraldo
Alckmin, the freshly appointed governor of São Paulo, refused
to declare a state of emergency and officially proclaim a drought.
This meant that São Paulo's (exemplary) laws on water regulation, designed for times of crisis, couldn't be enacted; since the
legislation was relatively new at this point and responsibilities
were still being questioned, it couldn't take effect without an
official declaration of crisis. Instead, nothing happened. Water
consumption wasn't regulated, and as the drought progressed
it was (as always) mainly the people in the slums who were left
with no clean drinking water and outbreaks of cholera.[26] It was
only after the crisis, in which—as the *Guardian* reported in
January 2015—"the taps ha[d] run dry and the lights ha[d] gone
out,"[27] that São Paulo invested in its water infrastructure and
made sure that going forward, local authorities can respond
to the drought conditions with which they are confronted.

For a positive example, we can look to China. In several
extremely fast-growing urban areas, water infrastructure keeps
pace with population growth and changes in rainfall. But the fact
that China gets better results than some democracies doesn't
mean that dictatorships might deal better with climate change.
The residents of Chinese cities are lucky that it is currently in
the state's interests to invest in adaptation measures—and that
these measures work. As the Covid-19 pandemic has clearly
shown, dictatorships are unwilling to admit their mistakes or
to learn if the situation changes—something for which the Chinese economy and people are still paying heavily. Adaptations
that work in the long term develop gradually and piecemeal,

often requiring great flexibility to respond to ever-changing situations. Forward planning is difficult in general and particularly when it comes to infrastructure, which doesn't appear overnight; managers have to make investment decisions long before a dramatic drought occurs. Democracies require the support of the population, and for good reason. But when people react with hostility to science-based decisions due to decades of disinformation, such as they did in Australia (see chapter 7), decision-makers need to plan well and communicate even better. And as we will see when we look at the disastrous flooding in the Ahr Valley in 2021, Germany too has a lot to learn (see chapter 8). Ultimately, these examples all show just how important truly independent media are, along with well-informed journalists and personnel in all departments. There's still a lot to be done here, all around the world.

It's not enough for hydrologists to have a seat at the table. For anything to change, political processes need to change on a fundamental level. Politicians of all levels of responsibilities need to know and understand the recommendations of climate specialists. Without expert advice, they can't make meaningful decisions, let alone talk with other stakeholders—especially the media—about why a decision makes sense and what needs to be tackled specifically. Our societies benefit greatly when hydrologists and other scientists consistently work with experts from the humanities and social sciences, particularly experts in economics, politics, and law. Together they can develop water management instruments that decision-makers and the public can understand. An important starting point in this process would be to clarify responsibilities; in both the heat wave in northwestern North America and in South Africa's water management, a lack of clarity on which measures are to be initiated when and by whom guaranteed that measures were enacted too late.

To ensure this doesn't happen, it's important to better understand the time factor—often it is crucial when planning decisions are made or whether those responsible immediately intervene. Ad hoc responses, as demonstrated by Cape Town in 2018 during the water crisis, are costly and unjust in equal measure—the price of water rose by a quarter in Cape Town during the crisis, disproportionately affecting poorer households. Irrigation bans may have delayed Day Zero, but they were also a disaster, mostly for those whose livelihoods depend on agriculture.

HUNGER

Cape Town may have avoided Day Zero in 2018, but that wasn't the case for the people in the surrounding fields. Thirty thousand jobs were lost, and the disaster destroyed many smallholder farms: at least 6 percent of the around 9,500 smaller agricultural businesses are never expected to recover.[28]

In addition to financial losses, the water shortage had significant social repercussions. Among smallholder farms in particular, illegal irrigation and cattle thefts led to conflicts that could lay the foundation for long-term social unrest if support measures aren't offered to all citizens.[29] The Western Cape is comparatively rich and the drought may not have directly led to permanently empty stomachs, but the reduction in exports to other parts of the world had some dramatic consequences for food security (whether people had enough calories and nutrients for a healthy lifestyle and whether or not the situation will lead to acute, life-threatening hunger).

Around two billion people worldwide regularly suffer from a shortage of good food. Fluctuations in the weather play a crucial role here, influencing around a third of unstable agricultural yields and affecting global food prices. There is, therefore, a direct connection between weather fluctuations and good, sufficient food.[30]

Dry periods such as those experienced by the whole of South Africa in 2007 and Cape Town from 2015 to 2017 put the food industry under particular pressure. When food becomes scarce and prices skyrocket, poor consumers often feel the impact disproportionately because they spend more of their household budget on staple foods. In Cape Town, this means that people living in the townships—the slums at the edges of the city—have to choose between water and food, an untenable situation. Often, parents will try to buy as much food as they can to ensure that their children can eat at least, which sometimes leaves no money to pay the water bill. So they get water from rivers, water that often isn't clean enough even after boiling, making disease unavoidable.

In a globalized world, these sorts of effects aren't limited to specific regions or countries. This is sometimes easier and sometimes harder to see. For example, the Covid-19 pandemic and the war in Ukraine (and the associated loss of Ukrainian wheat from the global market) directly influenced global food prices. In February 2023, more than 200 million additional people were affected by acute malnutrition or didn't know where their next meal was going to come from.[31] It is much more difficult to predict how climate change influences prices and food security globally—apart from cases extremely dependent on the weather, such as in Lesotho (discussed later in this chapter). This is because climate shocks come up against an incredible number of factors that determine local and regional food; these may include functioning irrigation (or lack thereof), various trading dependencies, crop diversity, and social and demographic factors. But because climate change affects all these aspects simultaneously when it comes to food, it also amplifies the various inequalities embedded within them.

When climate change hits locally, hunger and thirst often increase globally. The extent of this can be seen in the extreme heat in South Asia in 2022. The heat wave hit at a critical time, right in the final phase of the season for winter crops like wheat and barley, which were just about to be harvested—and therefore very sensitive. Likewise, summer crops such as legumes, coarse grains, oil seeds, vegetables, and fruits had just been planted and needed water and mild temperatures, not aridity and brutal heat.[32]

In India, the agricultural sector is one of the most important industries, employing almost two-thirds of the population (60 percent). The northern states of Punjab and Haryana manage a quarter of India's total wheat production, up to a third of which is estimated to have been destroyed due to the heat wave (10–35 percent). This made food much more expensive in India; prices rose by up to 15 percent in some regions. The impact was huge in Pakistan too, where around 40 percent of people work in the agricultural sector.

India is one of the world's biggest wheat producers—in 2023, it ranked second behind China—but its produce is mainly sold on domestic markets. When the wheat crisis intensified due to the war in Ukraine and constant rises in fertilizer prices, India decided (as mentioned in the previous chapter) to change its previous strategy and start exporting wheat. But once domestic food prices rose by 40 percent in March 2022—the highest they had ever been—the Indian government decided to ban wheat exports. This exacerbated several crises simultaneously on the global food market: the heat crisis, the wheat crisis, and the price crisis, a classic situation for an extreme event composed of multiple factors. But the true extent of the impact on the Indian and Pakistani agricultural sectors will only become apparent in the coming years, given that the economic effects of extreme heat are not yet being systematically investigated.

In other words, nobody knows how much climate change really costs.

What is clear, however, is that those who pay are the ones who can least afford it. And this applies not just to heat in India, but naturally also to droughts in South Africa.

THE COST OF CLIMATE CHANGE

The main reason that we don't know the true cost of climate change is that, as before, no efforts are being made to count or systematically estimate the damage. But there are clear indications that the price is much higher than we think. The most impressive example I'm aware of is Hurricane Harvey, which laid waste to Houston, Texas, in 2017 and caused estimated damage of 90 billion U.S. dollars.[33] Climate scientists and economists wanted to know how much of this was to be attributed to climate change. To find out, they used sources such as studies completed at the time, which showed that the rainfall brought about by Harvey (which caused a large proportion of the damage) was made around three times more likely by climate change. On this basis, they calculated that of those 90 billion U.S. dollars, a whole 67 billion dollars was to be attributed to climate change. Without climate change, three-quarters of the damage wouldn't have happened.

For comparison, a model developed by U.S. Nobel laureate William Nordhaus, which is used to calculate roughly how much climate change will cost the economy as a whole, calculated that 20 million U.S. dollars of damage would be incurred for the U.S. for the whole of 2017—not even a quarter of the damage that would later be caused by Hurricane Harvey. Why is there such a striking disparity with Nordhaus's numbers? Because extreme events are not considered in his or similar calculation models. These models, which remain the foundation for global and long-term economic deliberations, act as though only average weather exists. Word has now got out that this approach is subpar, but no fundamental changes have been made to the models as yet.[34]

Coming back to South Africa, we can use a study in which my team and I examined how climate change has influenced food insecurities in Lesotho.[35] In this study, we concentrated on a 2007 drought that proved dramatic for Lesotho. A constitutional monarchy, Lesotho was a British colony until 1966. Surrounded by South Africa on all sides, Lesotho is the only larger nation in the world to be entirely enclosed by another country (Vatican City and San Marino are too, but they're tiny by comparison). More than 3,280 feet (1,000 meters) above sea level, Lesotho has a very low rating on the Human Development Index (HDI), which measures a country's standing in terms of life expectancy, health, access to education, and an appropriate standard of living. In 2022, Lesotho was ranked 168 out of 193 countries and territories.[36]

In 2022, its residents had a life expectancy of just fifty-three years. Even today, Lesotho is battling with tuberculosis and has the second-highest HIV infection rate after Eswatini (a quarter of the population is affected by HIV).[37] Lesotho's agriculture is based largely on self-sufficiency. The country is mountainous, making it susceptible to droughts and flash floods, and (as mentioned already) it is largely dependent on South Africa, from which it imports more than three-quarters of its food (80 percent).[38]

The maize harvest in Lesotho and South Africa collapsed so significantly due to the 2007 drought that 400,000 people in Lesotho—around a fifth of the population—became reliant on food donations. Our analysis showed that a nonlinear reaction to climate change could be observed in Lesotho (as it was in Houston during Hurricane Harvey): even the comparatively small changes in the drought risk led to a disproportionately severe deterioration in food security.

It's not surprising that climate change increased the probability of a simultaneous drought in South Africa and Lesotho and, as we established, that it worsened food insecurity in Lesotho in combination with the changes in food production and prices that were to be expected. The climatic situation in both countries in 2007 was comparable with that of the Western Cape in 2015. However—and this needs to be made very clear—climate change did not cause Lesotho's food supply problems. As is so often the case, it made them worse. First of all, the main reason why Lesotho's food system is vulnerable to additional climate shocks is that Lesotho is highly reliant on regular, dependable rainfall and on a neighboring country with a similar drought risk. And Lesotho isn't the only one: in other Southern African countries, factors such as high import dependency, rain-fed agriculture, and great climate variability have a disproportionate impact on the standard of living compared with the rest of the world.

Lesotho, which the WHO anticipates will experience population growth of 26 percent by 2050 and—in view of climate change—is bracing for maize production to decline due to drought, has no choice but to adapt to climate change. This can be achieved, in part, by growing more of other grains alongside maize and finding other trading partners in addition to South Africa. But this is very difficult for a poor, landlocked country like Lesotho, as high transport costs seriously limit access to international grain markets. Even beyond this, Lesotho doesn't have an easy time of things. Its high HIV rate, an already fragile alpine landscape battling serious temperature fluctuations and highly irregular rainfall, and the very different effects of the Indian and Atlantic Oceans all present major challenges. And that's not to mention its limited budget for agricultural investments. To combat climate change, one of Lesotho's first steps definitely has to be tackling poverty. The consequences of the 2007 drought are so dramatic for Lesotho because they have lasted so long; they have forced half of farmers with smaller plots of agricultural land and large households out of self-sufficiency and into shortages, confronting them with severe hunger and malnutrition. As a result of the drought, people in Lesotho have become even poorer; poverty has increased by more than a third on average (37 percent). If you could afford three eggs before, now you can't even afford two.[39] Meanwhile, South Africa has come through the reduced exports of 2007 relatively unscathed. The difference in outcomes for Lesotho and South Africa echoes how differently people in Southern Africa are grappling with Day Zero.

THE RIGHT TO WATER

Water continues to flow from the faucets of Cape Town's houses (if they have any). But for decades, the city's poorest people have lacked adequate and reliable access to clean water. And the poorest people in Cape Town make up more than half its population. Most of them live in the Khayelitsha township, a sprawling, informal settlement. Most of their houses have been cobbled together from sheet metal, bits of wood, and cardboard, and the population is rising—in the last ten years alone, it has grown from 400,000 to 2.4 million people. Almost three-quarters of adults are unemployed there. A huge proportion of families in the township—almost 90 percent—are "moderately to severely food insecure."[40] More than half of the dwellings in Khayelitsha—which means "new home" in Xhosa—have no water supply, and the residents must get the water they need to drink, bathe, cook, and wash from the communal faucets. As the city administration rationed water more and more in 2018, hygiene in the settlement deteriorated drastically. There were more and more disease outbreaks because there was no water to wash hands or clean food. And since more and more water was being stored in contaminated containers or in unhygienic places, diseases transmitted through water (such as diarrhea, hepatitis A, and typhus) became major problems.

In other parts of Cape Town, the "problems" were quite different. Many people in the wealthy suburb of Rondebosch, for example, have swimming pools in their gardens. In times of drought, naturally they use their pools to store water for their gardens. It may be prohibited in Cape Town and the immediate drought area to use water to fill these pools, but it is possible to order tankers full of water from districts not affected by the restrictions—provided you have the money to do so.[41] And many people who do have the money turn to this option. For six thousand U.S. dollars, you can also have a well drilled in your garden, making yourself independent of the public supply.[42] So not everyone in Cape Town played their part in preventing the whole city from reaching Day Zero.

This would not just have been a moral imperative; everyone in Cape Town has the right to water, just like every other South African. Section 27 of the Constitution of South Africa confirms that all citizens have the right to water, which is also a human right.[43]

Human rights didn't take center stage in Cape Town's response to the water crisis, and even the research conducted immediately after the crisis concentrated largely on management and inequality—both still, unfortunately, topics in which human rights play no significant role worldwide. But it may only be a matter of time before that changes. The tide could turn as people increasingly recognize that courts may play a key role in international and national climate protection (see chapter 6). For example, given that the right to water is so explicitly enshrined in South African law, residents of the townships could sue the city government because they could not exercise their right to water during the drought. Such lawsuits may become very important in the future in clarifying what the right to water means in practice, how it can be better protected, and what role the different levels of government have to play in guaranteeing equal access to water during periods of rain and drought.

And this doesn't just apply to the right to water; other rights are under threat in South Africa, in part boosted by climate change. Many South Africans are aware of the constitutional right to housing, and they expect the government to ensure this is realized. South Africa does have programs for building social housing, but the homes—or the plots of land the cities can afford—are often so far out that it would be impossible to get to work or school.[44] And so a great many people continue to live in sprawling, tolerated townships that don't meet any building regulations, aren't connected to the sewage system, and don't have any waste disposal options. These townships don't exactly offer their residents the environment enshrined in the right to housing; naturally, this has striking consequences, particularly when people need protection against extreme weather. The Cape Town drought revealed many facets of inequality associated with the colonial-fossil narrative. Most of the people in Cape Town's townships, and most of those who lost their lives in the Durban floods mentioned at the start of this chapter, were people of color living in houses unworthy of the name.

5

POVERTY: THE ROOT OF THE CRISIS

Madagascar

"There is nothing to do here. I have no land so I cannot culti-vate anything. We live on wild tubers like fangitse and the red cactus in the forest. We sold all our domestic goods, including spoons... I have nothing left and it is painful."

"There is no activity, no work opportunity, no harvest, and noth-ing to put on the table."

"Without WFP's [World Food Programme] assistance, since we're left with nothing to eat, I think we would have died. I don't know who to turn to."

"We limit our consumption to fourteen handfuls of rice for one week."

"We save this food to ensure we have something in our tummy. We don't need to be full, just to eat something and to avoid being left with nothing to eat."[1]

LYTTON WASN'T the only place to suffer in the summer of 2021; the situation in southern Madagascar also deteriorated dramatically, although for very different reasons. If the United Nations World Food Programme (WFP) hadn't allocated additional funding, Tamaria and Tema (quoted above) and thousands more would have been at acute risk of starvation. Even now, their means of subsistence are far from secure.

For years now, Madagascar—particularly the south—has found itself in serious crisis. In stark contrast to the rainforest paradise many of us might picture thanks to the animated film, everyday life on the island is far from idyllic. In this region, 90 percent of the population (more than six million people) live below the poverty line. Following a prolonged drought in 2019–21, tens of thousands of people are suffering acute malnutrition and hunger to this day.[2] How could this happen?

THREE YEARS OF DROUGHT IN MADAGASCAR

In Androy, one of the regions most seriously affected in Le Grand Sud, the rainy season (December to February) normally provides half of the annual rainfall; just 20 percent of rain falls between April and September. But in the 2019–20 and 2020–21 seasons, rainfall only reached 60 percent of the average, leading to significant crop failures with dramatic consequences. Like South Africa, this is a region that depends completely on the rain, and the majority of the population live off subsistence agriculture (farming for personal use) and extensive pastoralism (natural pasture farming).

The rainy seasons had been similarly disappointing in 1990–92 and 2008–10. In the years between the extreme droughts, the region fought constant battles with hunger. Since 2017, the situation in Madagascar has been monitored by a coalition of various aid organizations and the Madagascan government. In 2019, 14 percent of the population—more than two million people—required urgent help with basic foodstuffs. By mid-2021, just two years later, this number had tripled. This was the first year that the food situation was classified as disastrous for tens of thousands of people, who didn't just lack nutrients due to unbalanced diets; they simply didn't get enough calories overall. Madagascar is one of the world's poorest countries and—at the time of this writing—is ranked 177 out of 193 countries on the Human Development Index (HDI), a little lower than Lesotho.[3] The HDI ranking is based on life expectancy, education, and household income. Due to positive developments in life expectancy and education, Madagascar's ranking has been improving slowly but surely since the 1990s. But poverty is and remains a permanent problem. In 2020, almost half of children aged five and under suffered from growth delays. The area most affected by the drought in the south of the country is also the poorest. Le Grand Sud's water and transportation infrastructure is extremely poor, and many roads are completely impassable during the rainy season, largely cutting the area off from the rest of the country; this increases prices in local businesses and makes it difficult for humanitarian aid to be delivered. Even in years with good or average rainfall, this regular isolation leads to high prices and malnutrition because key foodstuffs are too expensive or simply not available.

If the roads are impassable after drought years (or during brief, heavy rainfall in drought years), the almost-constant malnutrition quickly escalates into acute hunger. The poorest in society struggle to buy any food at all and must sometimes rely on gathering wild tubers and fruits—just like Tamaria and her family. In June 2021, FEWS NET, an NGO specializing in early drought warnings, calculated that crop plant production in southern Madagascar was 10–30 percent lower than the previous year and 50-70 percent lower than the five-year average. In other words, there was nothing to eat.

Madagascar's main rainy season—the months in which the roads are often impassable—is also the time at which the road network is needed most urgently. This is already a critical phase for most farmers, who rely on cultivating fields for their income. By this point, it's been a long time since the last harvest and food is getting scarce. If yields declined the previous year due to droughts, storms, locust plagues, or pests like the armyworm moth, there won't be much to put on the table during the rainy season. In bad years, people in this region may be forced to sell their possessions to buy food. But even this strategy only works if you can leave the area to visit markets where trade is still flourishing. In the southwest of Madagascar, herds of zebu cattle are kept as a form of insurance against drought. In the 2013-14 rainy season, for example, when lots of crops failed, cattle sales made up 56 percent of income.

The roads are clearly crucial to the region's resilience, and investment in the roads could have prevented at least some of the most dramatic effects of extreme weather events (and other shocks such as pest infestations). Despite this, there has been practically no investment in infrastructure in the last ten years, neither by the Madagascan government nor by NGOs or other development projects.

The impact of extreme weather events in Madagascar since 2020 has been heightened by long-standing problems such as poor transportation infrastructure and the lack of health-care and social systems. The high illiteracy rate also has a negative impact because people who can't read or write have very few income opportunities other than agriculture. The situation was further exacerbated during the Covid-19 pandemic. Unlike in Europe, where unemployment was prevented by generous state aid—or at least cushioned for the most part—Madagascar didn't put together any aid packages. If you lost your job in Le Grand Sud, you had to hope the Red Cross would give you food or go searching for wild-growing fruit. But to prevent the coronavirus from spreading, the government restricted freedom of movement, even in the regions most affected by drought and hunger. These measures drastically reduced employment opportunities for day laborers in particular and led to a downright exodus from the cities. Many people who had moved to the cities in search of work returned to rural areas and took on agricultural jobs. But they earned less than they would have in the city, and due to the increase in available workers, agricultural wages (already low) dropped further.

Jobs in the tourism, mining, and textile industries also disappeared during the pandemic. Strict health checks at the ports delayed food imports. Prices rose for imported goods, hitting the poorest demographic groups the hardest because, after poor harvest years, they rely heavily on imported rice.

These various factors—which have nothing to do with climate change and little to do with the weather—provide an insight into the complexities of establishing regional food security. Particularly in its southern regions, Madagascar has a huge investment backlog not just in its transportation infrastructure, but also in its educational and social systems. At the same time, the fight against corruption is stagnating and nobody is pushing for alternative income opportunities outside of agriculture—for example, by attracting companies to the region or offering adult education programs.

Even in years with no remarkable meteorological events, Le Grand Sud's semiarid climate is challenging for all forms of agribusiness and livestock production. Yields are only satisfactory in good years, which are becoming increasingly rare due to how the land is used. Extensive deforestation[4] has significantly reduced the ground's capacity to store water. This means that water evaporates faster, and the ground needs more or (even better) more regular rain to absorb the same amount of moisture. The state of the ground has a negative impact on pasture options because plants grow weaker roots and can't cope with dry periods. The entire system becomes more and more susceptible to disruptions such as drought years, extreme rain, and pest infestations.

We see similarly tricky situations in other parts of Africa too. In Kenya, for example, which also has a high risk of drought, an average drought (such as that of 2016–17) happens every five years or so, bringing devastating consequences despite its predictability. The meteorological conditions are tough: only a fifth of the country has regular rainfall. Over a quarter of the population (around 15 million people) and more than half of all livestock in Kenya live in arid and semiarid regions. People in rural areas battle similar problems as the residents of southern Madagascar. Kenya's low-income city dwellers are also vulnerable to climate extremes, as they often live in informal settlements with limited access to electricity, water, and sanitary facilities. Somalia (in East Africa) is even more vulnerable, as it doesn't just have regular periods of drought. It is also suffering civil war conditions, and for decades no government has managed to restore peace.

What we need to realize is that these challenges are not natural, nor are they God-given; they have developed over time.

COLONIALISM

In 2022, the UN World Food Programme—which was launched in 1961—declared the first ever "climate change famine" in Madagascar,[5] making it known that human-caused climate change is to be regarded as the major reason for food scarcity in the region. Since then, the WFP has supported the people of southern Madagascar. This is understandable from a humanitarian perspective, but scientifically speaking, the underlying diagnosis isn't easily substantiated. The attribution study conducted by my team in collaboration with the Madagascan weather service and Madagascan scientists confirms the IPCC's findings that the drought events that led to famine in Madagascar in 2021 can't be attributed to climate change.[6] Climate change does not play a statistically significant role in the frequency or intensity of periods of low rainfall in recent decades. But like many African countries, Madagascar also lacks weather records and data. This means we can't rule out the possibility that climate change has exerted some influence, but this can't be verified. Hunger and malnutrition in southern Madagascar have less to do with the arid climate and more to do with British missionaries and French colonialists.

After brutal wars against the Madagascan people, France occupied the island in 1895. Before this, communal Madagascan life was organized in various forms—from the kingdom down to locally organized communities with no overarching political structure. Under French colonial rule, the island formed a cohesive state for the first time, and while this did have some benefits, there were many drawbacks that can be seen to this day. Its state structures and institutions are largely a legacy of this era and have evolved very differently in different regions. The Madagascan judicial system is modeled on the French system and organized accordingly, but there are also traditional courts (dina), which rule on certain civil law disputes. The settlements and cities founded under or taken over by French colonial policy have a slightly lower level of poverty today compared with other areas, but they have particularly high levels of crime, vigilante justice, and resulting social conflicts.[7] Meanwhile, the regions settled by British missionaries have above-average rates of poverty. What both have in common is poor collaboration between general government and regional organizational forms; in times of crisis in particular, this means that none of the political authorities feel responsible for providing effective help to the people—a legacy of the colonial state-building project, which was based on enforcement and provided no opportunity for civilians to participate in politics. For Madagascar, and for many other African former colonies such as Uganda and Ghana, this means that promising reforms (such as the introduction of public health care and social security networks promoted by the missionaries) met with resistance from the people because state institutions lacked legitimacy at a local level.[8] These measures also disregarded the needs of the islanders. For example, the health-care system was based on Western concepts of care such as isolation and hospitalization, in stark contrast to the approaches that many community-oriented countries take to convalescence. Madagascar's many structural problems are founded in the same colonial-fossil politics of the former occupying powers that also led to climate change. In this respect, it isn't climate change directly that determines the dramatic effects of drought in Madagascar, but colonial politics are the cause of both phenomena.

So it isn't totally wrong to call the Madagascan food crisis a climate change famine. This does at least draw public attention to the situation on the ground and makes Western institutions and citizens more willing to donate; this includes highlighting where responsibility lies. Given its colonial past, the Western world does actually bear historical responsibility. But this doesn't mean the Madagascan authorities aren't responsible too. On the contrary, wider recognition of colonial and, above all, postcolonial blame must go hand in hand with a willingness to tackle corruption in the countries receiving reparation payments (see chapter 9). The money must not end up in the pockets of a few; it must be used to expand educational opportunities and disaster protection.

You could now argue that it actually makes no difference why we in the Global North bear responsibility for the situation in Madagascar, be it our historical responsibility as a colonial power, be it our current power as a global region that continues to drive climate change unchecked. If the only goal were to make people more willing to donate, then the reason for our responsibility is more effective if it foregrounds climate change—I'll come back to this shortly. But this question is about far more than that. This is a question of how we can shape a world that deals with the climate crisis so that as few people as possible will suffer. To achieve this, it is important to precisely identify the effects of climate change and make a distinction between this and other causes of disasters.

VULNERABILITY

Along with two colleagues, I developed this argument for a Nature Portfolio journal in early 2022. The journal gave our short text the title "Stop blaming the climate for disasters,"[9] which led to misunderstandings as we were accused of downplaying climate change. This is absurd; the whole point of my work is to research the impact of climate change, which is why I'm able to tell whether or not it's the main cause of a specific disaster. But making climate change the sole culprit puts responsibility on a vague higher power and declares it to be an abstract phenomenon against which authorities, organizations, and nations are seemingly powerless—a rhetoric we have already seen far too often in the context of climate change. Climate change takes on the role traditionally played by Zeus, Thanos, Zanahary, and other cosmic entities.

But human vulnerability is caused by human actions. This is apparent not just in the poorly thought-out urbanization processes described in the previous chapter, but also in any form of exclusion based on religion, caste, class, ethnicity, gender, or age.[10] These exclusions are based on systemic injustice, in which certain groups (such as women or people of other religions) are denied access to the resources they need, just like the women in The Gambia. Of course, the unequal treatment of women isn't limited to the African continent. In many Asian countries, women generally do not own the houses in which they live. This means they can't get loans or financial support from funding programs. After disasters, they struggle or are unable to rebuild their lives, making them even more vulnerable.[11] In Europe too, women are often worse off financially and thus more susceptible to the effects of extreme heat, for example.[12] Vulnerability is the product of social and political processes, and structural inequalities too are often created deliberately and enshrined in social and political structures. Blaming nature or climate change for disasters distracts from the processes that create injustice. Pointing to natural causes creates a politically convenient crisis narrative that rids us of responsibility.

This is dangerous; deflecting responsibility inevitably means that the status quo will continue, weakening the weakest even further. If you attribute disasters to nature, you pave a subtle escape route for all those responsible for vulnerability in the first place. No government, no city administration wants to hear that it is partly to blame for disasters. This applies to Canada, to South Africa, to Germany, and to Madagascar. But Madagascar is also the site of a peculiar alliance: the international NGOs and UN organizations active in the country say nothing about the failures of the government (which is working in extremely difficult conditions) in southern Madagascar because criticism would make the local situation even more chaotic. Plus, it's far easier to arrange Western donations with a narrative that clearly attributes responsibility to the West, the climate, and nature.

Climate change and the associated responsibility of the West is an obvious argument for donors. The story is plausible, easy to tell, and easy to understand. If aid organizations were to use "colonialism" as their campaign battle cry, it would be much harder to get to people and their wallets because, even now, there is nowhere near enough acceptance of the responsibility stemming from colonialism.

I don't want to accuse Western NGOs of being consciously calculating, of deliberately exaggerating the role of climate change. There remains an urgent need to better understand climate change and its consequences, to communicate the findings and, above all, to fight the causes. And yet I find it problematic that NGOs don't reference colonial responsibility in their arguments. If they did, they would have to scrutinize their own practices as well—and would then realize that many emergency aid delivery processes have colonialist structures themselves, at times upholding existing problems.

COLONIALISM AND RACISM IN NGOS

After most colonized nations achieved independence in the 1960s, the Global North was eager to make the newly created states dependent on exports. Aid was explicitly used to reduce Western food surpluses. Hollis B. Chenery, who worked for the U.S. Agency for International Development (USAID), stated that "the main objective of foreign assistance, as of many other tools of foreign policy, is to produce the kind of political and economic environment in the world in which the United States can best pursue its own social goals."[13] In hindsight, this also applies to projects that went beyond food deliveries. Development aid is often aimed at large-scale projects that tend to do more harm than good. For example, agricultural operations were funded that put excessive strain on the soil with machines and chemicals because this allowed test fields and sales markets to be created for Western products. Their comparatively high yields worsened the situation for small farmers to such an extent that they became reliant on food aid.

And as with most infrastructure measures, many of the local people affected didn't benefit in any way. Irrigation systems were installed in the Sahel region, but often not where the farmers actually needed them. One now-infamous example is around the city of Richard Toll in Senegal, where a great deal of pasture land was rendered permanently unusable in favor of irrigating sugarcane plantations for the benefit of a handful of land barons—with no compensation whatsoever for the farmers.[14] In 2016, the Food and Agriculture Organization of the United Nations (FAO) estimated that just 54 percent of the irrigation systems funded by the Sahel programs were actually used. And if production was increased beyond what the farmers needed for personal use, their situation still didn't improve for the most part because the markets were so far away that they often couldn't sell their grains and vegetables before they spoiled. Dam construction, financed by various aid organizations, had similar problems. In many cases, the land downstream was simply used as a flood plain, destroying local farmers' farmland and pastures without providing replacements.

Western institutions, particularly the World Bank, consciously decided that the newly created nations would bear the costs of overcoming colonial underdevelopment and had to pay back the loans for the failed developments of the 1950s and 1960s. It wasn't until 1996 that the International Monetary Fund (IMF) and the World Bank floated the idea of debt relief for "developing countries." Long-standing financial dependence restricted the ability of new nations to invest in their own, tailored infrastructure projects. Development aid has therefore also played a part in perpetuating poverty and dependency in the regions it is supposed to be helping. Colonialism isn't just a legacy of the European colonial powers; in the form of post-colonialism, it has caused and is still causing further damage for which today's Western powers must take responsibility.

After many of the major infrastructure projects of the 1960s failed in a similar way to those in Richard Toll, in the 1980s the World Bank began to finance smaller projects. Dams of smaller dimensions, local roadbuilding, and other infrastructure projects were funded, but often without success—the large-scale irrigation projects had forced many people out of their villages because they were either underwater or located by dried-out rivers and fields rendered unusable. How were people supposed to survive in conditions like this? Ultimately, all these measures achieved overall was greater dependency on international aid, which in recent decades has shaped our Western image of a poor, "underdeveloped" African continent. The Covid-19 pandemic and the Russian war of aggression amplified this problem again. Infrastructure projects (of any size) are no solution to improve the situation in the long term; what is required is to build functioning local markets at which subsistence farmers can make a profit by selling the grains, vegetables, and fish they don't need for personal use. Alternative income opportunities must also be developed in the small business and service provision sector for those who can't live off agriculture. This can be achieved, for example, through the awarding of microloans that people can use to set up small businesses, establish themselves as service providers or medium-sized businesses, or pursue further education. Mechanisms are also required to prevent large tracts of arable land being exploited by foreign investors for the cultivation of fruit for export.

For all these measures to be implemented, certain conditions must be fulfilled. Regional grain storage must be established so that farmers are no longer forced to sell straight after harvesting. This could help to stabilize prices. A better understanding would also be required of optimal crop rotation for different types of soil. Communication between farmers, veterinarians, transport companies, and markets should be optimized too. This can only be done with the participation of the people whose lives are affected. But most international aid projects haven't managed this yet. In southern Madagascar, NGOs organized food deliveries when the situation was particularly bad, but there was no structural support to improve the situation in the future and reduce dependencies.

Why am I talking about all this when my book is supposed to be about climate justice and the 2021 drought had nothing to do with climate change? Because Madagascar is a particularly clear example of how closely adaptation is connected to justice. Vast areas of Madagascar are not adapted to the Madagascan climate. This is a justice issue rooted not least in colonialism. Although the age of colonialism is historically often considered to have ended, it lives on in power structures. This becomes clear in international climate negotiations too (more on this in chapter 9).

Less obvious, but at least as problematic, are colonialist and racist structures in NGOs and health organizations. In an article focusing on international development aid, global health expert Lazenya Weekes-Richemond states very clearly what needs to change to actually reduce vulnerability in Global South countries. In many cases, white women hold leadership positions in regional offices. Weekes-Richemond makes it crystal clear that development and collaboration can only function successfully if the causes of inequality are addressed. White women sometimes do more harm than good if they throw their weight behind the fight against sexism while consolidating racist structures:

"What the sector so desperately needs is active reflection and action from white women to interrupt the harmful habits and dismantle the structures that perpetuate white supremacy in day-to-day work."[15]

As always, development projects in the Global South are largely managed by white people. Mostly by women—who of course aren't usually actively trying to cause harm, but do so in practice by failing to reflect sufficiently on their own horizons of experience and education; as a result, the interventions they initiate aren't tailored to local needs and therefore often cannot improve living conditions in the long term.

Weekes-Richemond calls on white women to be true allies who are aware of their privileges and consciously use them to help people in crisis regions. If they did, then women of color wouldn't have to fight both them and the patriarchy simultaneously. Leaders in particular should scrutinize their prejudices and motivations and work with women of color on an equal footing. Simply put, this means paying them fairly, being willing to learn from them, ceding leadership positions to them, and, above all, stepping aside and making space for men and women from the region in question. The Global South doesn't need white saviors; it needs sustainable development aid. In the words of Lazenya Weekes-Richemond: "Millions of lives are at stake and it is high time that we put our heads together and do the uncomfortable work of working ourselves out of a job."

BACK TO THE CLIMATE

The 2019–21 drought in Madagascar had nothing much to do with climate change. But this doesn't mean that climate change isn't relevant to the region—quite the opposite. While droughts and locust plagues have been putting strain on the ailing social and infrastructure system for decades, climate change is adding

new shocks. As in the whole of Africa (see chapter 3), temperatures are rising considerably in Madagascar too, and periods of
extreme heat deviating from the median by up to 2°C (3.6°F)—a
lot for this latitude—have increased significantly.[16] Madagascar
has always been hit by cyclones which, due to climate change,
are accompanied by stronger rainfall and cause greater damage.[17]
So we definitely shouldn't ignore climate change in the context
of Madagascar's droughts. It may not be a primary cause, but it's
certainly an additional threat.

It's not easy to determine the extent to which generally low
resilience to shocks is connected to political conflicts, wars,
and civil wars on the one hand, and to natural and climate-
change-related changes and fluctuations in the weather on the
other. We also need to ask ourselves whether there might be a
causal relationship between changes in weather and conflicts,
and between changes in weather and migration, either within a
country or across borders too.

When extreme weather events become more frequent and
intense, the people affected have less and less time to gather
themselves and recover. In the long run, the situation deteriorates for the local population—sometimes leading to political
unrest, which heightens instability and inequality. Against the
backdrop of changes in the climate, social tensions frequently
intensify and ethnic polarization flares up repeatedly. Extremist
groups use these circumstances to intensify conflicts and attract
followers. The Sahel region is a tragic example of how existing ethnic and religious tensions lead to armed conflicts in the
context of extreme weather. Extreme weather events such as
flooding have made the already difficult security situation even
worse by damaging public and private infrastructure and wip-
ing out entire harvests and livestock herds. There are now more
than 2.2 million internally displaced persons in the Sahel region.
In addition, almost 900,000 refugees from other crisis areas

nearby seek sanctuary there. The interplay of socioeconomic vulnerability, consequences of extreme weather, and simmering ethnic conflicts leads to violent confrontations between farmers and nomadic herders in many parts of the Sahel region. After years of ethnic discrimination and a lack of political participation, many nomadic herders allow themselves to be recruited by extremist groups such as Ansarul Islam in Burkina Faso or Ansar Dine in Mali.[18] In this way, climate change amplifies inequality without directly affecting the droughts in the Sahel region.

The ethnic fragmentation in this region exacerbates matters further; for example, if people will only help neighbors of the same ethnicity. Patients are turned away from hospitals if they don't belong to the region's dominant group.

Despite mounting evidence connecting climatic conditions with conflicts, explanations are often unsatisfactory. Studies often confuse correlation (a simple mutual relationship) and causality as they relate to the links between climate change and conflicts.[19] Comprehensive theories are often so complex that it's difficult to apply them to the quantitative data available. All in all, research into the causal mechanisms that link climate change with conflicts has so far proved inadequate.

The same goes for the second major topic that keeps on popping up, particularly in connection with Africa: migration. Immigration to Europe or the U.S. from Africa or Central and South America as a consequence of climate change is often portrayed as a nightmare scenario by nationalist and conservative politicians and used by more progressive politicians as a warning to approve ambitious climate protection targets. It's doubtful whether this scenario will ever actually come to pass; so far, there has been no transcontinental migration that can be unequivocally attributed to climate change. In the scientific world, it is beyond doubt that climate change will alter migration, but here too research remains thin on the ground.

Political and social scientists may have been tackling this topic for a while, but it's still relatively new for natural scientists and even meteorologists. And so we lack empirical foundations on which to identify causal relationships. Plus there's the fact that the conditions for migration (and its many different forms) are enormously complex. Most people who leave their homes stay within their own country and often only leave for a few months or years. International migration works in a similar way; most people stay within a region and return in the foreseeable future. Fleeing and emigrating are fairly rare in comparison, even if populist voices in Europe and the U.S. like to claim otherwise.

One of the major problems that led to the dramatic food situation in southern Madagascar in the drought years is the lack of mobility for the people who live there. This shows why the inequality heightened by climate change doesn't necessarily lead to more migration, as is often assumed: growing poverty keeps the people affected stuck where they are. In fact, fewer people migrated during the drought period, which made conditions in southern Madagascar even worse. The lack of migration opportunities must therefore be seen as a consequence of crises, which in turn exacerbate the lack of opportunities.

Anyone who manages to emigrate will escape the extreme weather events and other crises in the area. If emigration isn't part of specific plans for adaptation to climate change, people are more likely to stay trapped in those parts of the world where vulnerability is particularly high: more than 38 percent of the global population live in comparatively dry climate zones and are disproportionately affected by climate change because their resilience to even small changes is already incredibly low.[20]

Migration decisions are motivated by a range of factors that may be economic, social, political, ecological, and not least cultural in nature. Climate change interacts with these factors and also influences a person's decision on whether to leave their homeland. As I've already stated, it hasn't been proven whether this interaction leads to more migration across borders overall.

Several studies have been published in recent years on the manifold links between environmental changes, social factors, and migration that focus mainly on African dry zones south of the Sahara. Climate change and extreme weather events play a central role in these studies as a driver of various forms of migration. According to these studies, 23 percent of the 9,700 rural African households surveyed migrated due to onerous environmental conditions, but most of them stayed within the region. However, these studies also show that the decision to migrate often depends on factors that are difficult to objectify, such as personal networks—whether a person has close contacts in other regions, cities, or (rarely) countries that would make it easier to start afresh somewhere new. No precise conclusions can be drawn from these studies. But there are two clear findings: Firstly, climate change will not provoke mass migration from Africa to Europe or on other continents, because other factors have much more influence over such serious life choices. Social and political structures determine whether a region is and remains habitable. Secondly, migration and displacement are expected to increase strongly within individual countries. One important reason for this is that the people most affected by climate change in Africa do not have the financial resources and social networks required to flee to another continent.

The latest research also makes clear that changes in the weather related to climate change are rarely a direct reason for migration; this tends to be inspired by indirect factors such as health risks, which are rising significantly due to climate change. Higher temperatures are increasing the spread of diseases like malaria. The dramatic rise in heat waves reduces productivity (see chapter 3) and puts immense stress on the cardiovascular system, leading to secondary diseases. This could cause people to flee from the countryside to the city, despite its even worse heat conditions and air quality. The poorest would-be migrants would probably put up with this, heightening existing difficulties, such as a lack of job opportunities and adequate living space in cities due to the rapidly growing population.

UNDERSTANDING CAUSES

The inequality established over the centuries by colonial rule, politics, economics, and development aid means that today, in the first decades of the twenty-first century, the accounts of Tamaria and Tema, the two women quoted at the start of this chapter, read like something out of Hans Christian Andersen. But their story is no fairy tale, nor is it a bleak look into the future of climate change; it is the bitter reality of the extremely unequal world we have created. We can change it, but to do so we have to understand the past and name the fundamental causes of existing inequality. Only by learning to distinguish between climate change and other causes can we adapt effectively. This often takes some getting used to, particularly for white Europeans and North Americans—we need to change the way we think, restructure science, development cooperation, and activism, and allow for true diversity. But it's certainly not impossible.

PART III

FIRE

How Climate Litigation Is Pushing Back Against Disinformation

6

THE END OF THE RAINFOREST

Brazil

"The forest is our home; it heals our soul. Without it, we are nothing."[1]

"If the Amazon suffers, the world suffers."[2]

"Loggers are extremely dangerous."

"We shouldn't be doing it. It's the duty of the federal and state governments, but since they are not protecting it now, we are the ones doing it."

"It's disturbing to see the state inciting threats against the state itself."

"'I want to make a deal with you to sell timber,' the caller said. 'If you don't accept, you will die.'"[3]

IN JULY AND AUGUST 2019, as Madagascar battled the onset of its three-year drought, a record number of fires—almost 20,000—drew the world's attention to the Brazilian rainforest. The rainforest often burns in summer—particularly in hot, dry years—and particularly dramatically in 2015, then the hottest year ever measured at 0.91°C (1.64°F) above the hundred-year average (1901–2000). Eventually, 2019 became the second-hottest year with an anomaly of 0.97°C (1.75°F), although this is not unusually warm or dry in a world with 1.2°C (2.16°F) of global warming; 2019 would later be overtaken by 2023. What made these extreme fires all the more worrying was that there was no obvious cause. The public outcry extended across many countries; the G7 and even Pope Francis called on Jair Bolsonaro, then president of Brazil, to do everything in his power to contain the fires as quickly as possible.

But rather than averting the looming disaster and saving the rainforest and its inhabitants from the flames, Bolsonaro used his international platform to stoke nationalist resentment. Without even the slightest evidence, he blamed international NGOs for the fires. Responding to the appeals of the G7, he declared that "there could be... I'm not affirming it, criminal action by these 'NGOers' to call attention against my person, against the government of Brazil. This is the war that we are facing."[4]

This response in particular drew international attention to Bolsonaro's role in the Amazon fires. There can be no doubt that his policies have caused great damage that extends far beyond Brazil. The attempts of the G7 and the pope to exert political and moral pressure did not prove successful. Powerless to stop Bolsonaro's politics and thus the flames, some groups tried a different tack and asked whether the president could be sued.

But let's not get ahead of ourselves. What exactly happened in 2019?

THE FIRES OF 2019

Between January and April 2019, satellite images showed an unusually high number of fire sources in the Amazon, particularly in regions where lots of illegal logging had been reported. By August, smoke pollution was so high that several Brazilian states had enduringly dark skies, and on August 19 the smoke began to spread southeast, farther and farther away from the Amazon. When the sun failed to rise in the city of São Paulo, the Brazilian and international media began to ask whether the fires were connected to illegal logging, Bolsonaro's environmental policies, and climate change.

First, the scientists at NASA[5] tried to find answers by interpreting the satellite images that had told the world about the fires in the first place. The data required to localize fire sources is generally available, but researching the causes takes time, and it wasn't until six months after the fires that NASA completed its assessment of the 2019 Amazon fire season.

"There is no question the 2019 Amazon fires were unusual, but they were unusual in specific areas and ways," said Douglas Morton, chief of the Biospheric Sciences Laboratory at NASA's Goddard Space Flight Center. "Fortunately, we did not see forest fires burning uncontrolled through the rainforest like we have during past drought years [above all, 2015]. What we did see was a worrisome increase in deforestation fires in certain parts of Brazil."[6]

But while the number of fires broke records, the destruction of the rainforest was limited because 2019 was not a drought year. "The real nightmare scenario would have been deforestation fires at the level we had in 2019 *during* a drought year," explained Alberto Setzer, a senior scientist at Brazil's National Institute for Space Research (Instituto Nacional de Pesquisas

Espaciais, or INPE). "You would have seen fires spreading into the rainforest and burning unchecked for months."[7]

NASA's satellite images (generated with the help of MODIS, a measurement and orbit instrument with imaging functions) are freely accessible online.[8] But it isn't easy to interpret the fire sources, which are marked with red dots, because small, seasonal agricultural fires look the same as larger bonfires or grass and brush fires, which can quickly run rampant in dried-out wetlands or bush savanna. So we can't really tell from the satellite images alone whether 2019 was a notable year in this respect. When we compare the data collected over the years, the MODIS observations for the initial years in particular (2001–2004) show extremely high fire activity; it wasn't until 2004 that Brazil enacted a series of environmental regulations aimed primarily at reducing slash-and-burn agriculture and other forms of deforestation in the rainforest.

However, these regulations were barely enforced, especially once Bolsonaro took office in January 2019, and deforestation once again increased considerably. But the true scope is difficult to determine because illegal loggers often try to outsmart the satellite system—for example, narrow strips logged at the edge of the rainforest are difficult to spot from the satellite data.

In 2019, slash-and-burn agriculture began in the dry season, and this was one of the main reasons for the international attention. When public outrage reached its peak in August and September, the Brazilian government sent tens of thousands of soldiers into the rainforest to fight the fires, despite Bolsonaro's initial ignorance. Additional heavy rainfall also dampened fire activity in large swaths of the Amazon, meaning that the number of fires and burnt areas was ultimately below average in many regions. After a dramatic start with an extremely high number of fires, a major disaster was averted.

According to Setzer, "[i]f the military hadn't intervened and the rains hadn't picked up, there is no doubt that the totals for the year would have been much higher."[9]

BOLSONARO AND ENVIRONMENTAL PROTECTION

The Amazon rainforest is the Earth's green lung, and 60 percent of it is located in Brazil. This means that Brazil has the world's largest virgin forest, but also loses the most forest each year—15 million hectares (3.7 million acres) disappeared in 2021 alone,[10] an area around the size of Connecticut. This isn't just a dramatic loss of biodiversity; it threatens the existence of many Indigenous groups and amplifies climate change because less CO_2 can be stored as the forest disappears. Deforestation is Brazil's greatest contribution to the acceleration of climate change, which makes it so much more important to maintain the forest it still has, which will help stem global warming. In the 2015 Paris Agreement, Brazil pledged to stop all forms of illegal deforestation (which makes up more than 90 percent of all deforestation[11]) in the Amazon region by 2030.

To meet these obligations, the main thing the Brazilian government needs to do is fight the criminal groups responsible for a majority of the logging. Luiz Inácio Lula da Silva (or "Lula"), Brazil's current president, has now promised to revive the laws enacted before Bolsonaro's term of office and end illegal deforestation. But the statistics show the difficulty in rolling back Bolsonaro's policies: in February 2023, two months after Lula took office, deforestation was 62 percent higher than the previous year. This is despite Lula working with Marina Silva, the same environmental minister who managed to reduce deforestation considerably during his first term of office (from 2003 to 2011). In 2023, Silva presented a plan to prevent illegal deforestation. In essence, the plan is to reactivate the National

Environment Council and the Amazon fund that Lula launched in 2008 to protect and conserve the rainforest. The National Environment Council (Conselho Nacional do Meio Ambiente, or CONAMA) is Brazil's highest deliberative body for environmental policy. It was established by Law no. 6.938/81, which also created the National Environmental Policy. CONAMA operates under the Ministry of the Environment and is responsible for establishing guidelines and standards for environmental protection and conservation, as well as for regulating activities that may impact the environment. CONAMA is composed of representatives from various government agencies, including federal, state, and municipal agencies, as well as representatives from civil society organizations, industry, and academia. It played a crucial role in combating deforestation, but all funding was cut off under Bolsonaro.

Bolsonaro clearly had different priorities during his time in office, actively hindering the implementation of laws to protect the Amazon and weakening the country's environmental authorities. He lambasted organizations and individuals who advocated for rainforest preservation and, according to environmental officers and local residents, gave the green light to the criminal networks involved in illegal logging. In doing so, he didn't just endanger the people who live in the Amazon region; he also undermined Brazil's efforts to uphold its commitment to reducing its greenhouse gas emissions. It will be some time before environmental agencies and forest conservation organizations are fully staffed again, especially since their employees suffer constant harassment from criminal gangs.

Bolsonaro's administration is a telling example of how misguided climate policy strengthens inequality, undermines human rights, and kills without conscience. The former president tolerated the outright murder of the people who wanted to protect the forest.[12] He also indirectly accepted human deaths because his deforestation policy directly boosted climate change and continues to cause huge damage to the Brazilian population to this day.

DEFORESTATION

Since records began in 1959, deforestation—particularly of tropical rainforests—has accounted for almost a fifth (19 percent) of global CO_2 emissions. The deforestation of the Amazon (or, to be more precise, the CO_2 released through the burning and decomposition of trees and the amount of CO_2 no longer extracted from the air by felled trees) is the greatest contributor to global climate change after the burning of fossil fuels.[13] To achieve the goals of the Paris Agreement and limit global warming to $1.5°C$ ($2.7°F$) above preindustrial levels, the emissions caused by deforestation also need to drop rapidly; according to the IPCC special report, most scenarios for minimizing emissions require all deforestation-related emissions to be stopped by 2030[14] so that the temperature target can be met. In other words, from 2030, no more trees may be felled unless they are directly and sustainably reforested.

But deforestation rates have risen considerably since Bolsonaro took office on January 1, 2019. In the first year of his administration, deforestation was higher than ever (34 percent more than 2018). The deforestation of the Brazilian Amazon accelerated further in 2020 and exceeded the previous year's level by another 10 percent: for every 100 trees felled in 2018, another 144 were lost. In a report[15] backing a lawsuit against Bolsonaro, the Bolsonaro administration is blamed for deforestation beyond the average rate of 2009–2018. To assess the scope of deforestation in the Amazon that is to be attributed to the Bolsonaro administration in 2021 and 2022, the report presents three scenarios: a "low" deforestation scenario that holds deforestation rates at the 2020 level, a "medium" scenario that continues the increase in deforestation observed between 2019–20, and a "high" deforestation scenario in which deforestation increases linearly to reach, in 2022, the peak levels observed twenty years before. The data available today suggests that the medium scenario is most likely, meaning that the mere rise in emissions caused by deforestation during the Bolsonaro administration makes up an estimated 1 percent of global CO_2 emissions each year, approximately equivalent to the total emissions of the United Kingdom. Climate change is also accelerating because the herds of cattle in cleared areas of forest emit harmful methane. The state of the rainforest at the end of Bolsonaro's term demonstrates the devastating impact of an individual's politics, in the longer term too, and how difficult it is to halt or reverse this trend.

Even in 2023, deforestation goes on and Bolsonaro's environmental policies continue to cast a long shadow. Radical and consistent political changes remain necessary to stop the destruction. As summer 2023 came to an end, there was cause for cautious hope[16] after logging decreased by 66 percent. It remains to be seen whether this can be maintained in the coming years. Lula definitely has some major challenges ahead of him.

CLIMATE DAMAGE

The 1 percent of global emissions caused by the Bolsonaro administration are also responsible for 1 percent of climate damage. In chapter 2, we looked at a heat wave that wouldn't have happened without climate change, and in which more than six hundred people died. One percent of this number, at least six lives, can be added to Bolsonaro's ledger. But the total number will be higher for a year like 2021, which saw further extreme weather events connected with human-caused climate change that led to many deaths: hurricanes in the U.S. and Bangladesh; flooding in Germany, India, and Nepal; and heat and forest fires in many parts of Europe and the U.S.[17] Nobody has actually counted these deaths, and many—as we saw in chapter 3—are never included in statistics anyway. But we can definitely assume that there were over 100,000 deaths. One percent of that is over a thousand deaths. The real figure is probably much higher—plus there's the damage to nature and infrastructure.

As we have already seen in chapter 2, heat-related deaths arc one of the most dramatic consequences of climate change. Based on a current estimate, within the next eighty years there will be over 180,000 heat-related deaths worldwide solely due to the 1 percent of additional emissions caused by Bolsonaro's policies.[18] That's 180,000 people who will lose their lives, even if global emissions decrease considerably. This estimate is very conservative and doesn't take into account anywhere near all the climate-related damage caused by these emissions.

FOREST FIRES

Although humans are usually to blame for the ignition of forest fires, the climate and weather can be key factors in how far fires spread and how intense they are. Drought periods are particularly good accelerants. While droughts occur randomly in the midlatitudes, in the tropics they are connected (at least in part) with fluctuations in the sea surface temperature.

These fluctuations are linked to a phenomenon known as the El Niño-Southern Oscillation (ENSO). El Niño and La Niña are fluctuations in the Earth's climate system that often have global consequences. Southern Oscillation is the changes in air pressure between the Tropical Eastern and the Western Pacific that accompany episodes of both El Niño and La Niña and influence factors such as trade winds, which in turn play a major role in the global climate system.

El Niño is a warming of the sea surface temperature that occurs every few years and is typically concentrated on the central-eastern equatorial Pacific, which means it can mainly be observed by the coast of countries like Peru. Major El Niño events last occurred in 2015–16 and 2023–24. We talk of an El Niño event when the sea temperatures in the Tropical Eastern Pacific are 0.5°C (0.9°F) above the long-term median, while La Niña indicates the opposite. These are episodes in which the average sea surface temperature in the equatorial Pacific becomes cooler and ultimately at least 3°C (5.4°F) below the average. The years 2020–2022 were unusual because they brought a three-year La Niña event that led, among other things, to droughts in Argentina and other South American regions key to global wheat and soy production.[19]

The ENSO phenomenon is the world's best-known and most influential natural fluctuation in the climate system. These fluctuations determine which sections of the world's oceans are warmer and which are cooler. Trade winds aren't the only thing that change with the water temperature; other dominant winds over the oceans are also affected, in turn influencing the development of low-pressure areas. This means that rainfall and droughts in the Amazon are partially dependent on the ENSO phenomenon too. So far, no research findings have indicated that human-caused climate change alters the frequency of these natural fluctuations. Ocean temperatures have risen by a total of 0.88°C (1.58°F) due to climate change, independent of the ENSO.[20]

To address the question of whether or how climate change influences forest fires in Brazil, I first need to make it clear that while weather plays an important role, it is not the only factor. "Fire weather" can't simply be equated with droughts because the risk of forest fires increases in general when temperatures are high, conditions are dry, and the wind is rising. These factors can be summarized in an index such as Canada's Fire Weather Index (FWI) or Australia's Forest Fire Danger Index. These indices combine the influences of temperature, humidity, and wind speed (and usually the rain quantities of the previous months) to determine the degree of fire risk. Not every index is suitable for every region; bushfires in Australia behave differently than fires in the Amazon, just as British moor fires play out differently than fires in German pine forests.

The Canadian FWI proved suitable for the Cerrado, a vast area of tropical wet savanna in southeastern Brazil, because it was the most precise way to demonstrate the scope of the forest fires that occurred in 2015. This index isn't limited to the current weather; it also includes previous dry periods. The question that attribution studies must now address is whether the effects of human-caused climate change have made higher FWI values (extreme risk of forest fires) occur more often. With the aid of observation data and simulations, the current climatic conditions can be compared with the conditions that would prevail if humans hadn't emitted additional greenhouse gases into the atmosphere. The results can be used to estimate the extent to which climate change influences the risk of forest fires. But this is simply an estimate; unlike with heat waves, for example, which can be measured directly using temperature values, the Fire Weather Index only states the probability of a forest fire and doesn't comment on the actual event.

As one study[21] shows, in the Cerrado and the "arc of deforestation" at the edge of the Amazon, human-caused climate change actually fosters the weather conditions for severe forest fires that have occurred much more frequently in both regions. An extremely high Fire Weather Index, which occurs around every ten years in today's climate, would appear once every twenty years without climate change, and once-in-a-century weather for forest fires would be expected once every five hundred years without climate change. Once again, as with heat, the more extreme the event, the greater the influence of climate change.

The change in the FWI in response to climate change can be explained by two main factors. The rise in temperatures and simultaneous drop in humidity across the Cerrado and the "arc of deforestation" ensure that the risk of forest fires has increased much more than in other parts of the world such as Australia, which is also afflicted by many dramatic bushfires (see chapter 7). The risk of forest fires also increases in El Niño conditions such as those of 2015 and 2023, and decreases slightly in La Niña years like 2022. Therefore, human-caused climate change and the El Niño episode increased the probability of severe fire weather in 2015. However, extreme forest fires also occurred in years when El Niño played no role, such as 2005 and 2010. So there is a connection, but not in the sense of simple causality; these phenomena are too complex for that.

Fire weather is just one aspect of forest-fire risk. Ultimately, actual fire events depend on the susceptibility of the local landscape, which in turn is influenced by how the land is used: untouched virgin forests burn worse than those where trees have already been felled, and young forests catch fire more easily than old. Climate change also plays a decisive role, as we can see in the Cerrado both in the rise in temperatures and in the increased frequency of droughts.

The connection between fire weather and climate change means that there are already various adaptation measures (such as improved drought monitoring and warning systems and more effective fire-protection measures and firefighting plans) to prevent land degradation, carbon emissions, and damage to economies and health. In principle, all of these measures can be implemented in the regions affected. To ignore them is to neglect human lives, as forest fires don't just endanger the environment, infrastructure, and real estate; they place people in immediate danger too. Although the number of direct deaths is usually lower than with other extreme events, forest fires are harmful to health. The smoke is made up of fine particles that can get deep into the lungs and then into the bloodstream. These may even be more toxic than the road-traffic particles that claim many victims (see chapter 2). These air pollutants definitely exacerbate existing respiratory disorders and can also have lasting effects on the cardiovascular system, and even on pregnancies. Plus there's the fact that forest fires often go hand in hand with heat, and the consequences of one will amplify the other. The residents of New York and other cities on the U.S. East Coast experienced this firsthand in May and June 2023, as the worst forest fires in Canadian history raged in Quebec. Even hundreds of miles away in the Big Apple, clouds of smoke made it difficult to breathe. Climate change and its far-reaching consequences made their presence felt in America's lungs.

So Bolsonaro's climate policies don't just have deadly consequences on account of increased emissions. Even after the change of government, defenders of the rainforest still risk life and limb whenever they go up against the loggers violating Brazil's environmental laws. Illegal deforestation in the Brazilian Amazon is largely performed by criminal networks with the logistical means to coordinate the large-scale felling, processing, and sale of wood, mainly to the U.S. and Europe. Some environmental officers refer to these armed gangs as the "Ipê Mafia," referring to the ipê tree, its wood some of the most valuable and coveted in the world. But the loggers cut down other tree species too, as the ultimate goal of their clients is to clear the forest completely to create space for livestock or agricultural crops. Cattle herds are the most popular choice for these newly acquired spaces, as the meat from the animals can be sold to global fast-food chains. Every burger and every steak contributes not just to the destruction of the rainforest, but also to climate change. As long as meat consumption remains high around the world, Lula's task of stopping the logging will remain extremely difficult.

For Brazil to fulfill its obligations from the Paris Agreement, the criminal gangs must be broken up; this would also protect the people who defend the forest—often Indigenous groups such as those quoted at the start of this chapter. It remains to be seen if Lula can succeed in overcoming the mess left behind by Bolsonaro and taking action against the criminal gangs.

BEFORE THE COURTS

Humans have the right to live in a climate-friendly environment. We can deduce this from the UN Report of the Special Rapporteur on Human Rights and the Environment, published in 2019, which states that "a failure to fulfil international climate change commitments is a prima facie violation of the State's obligations to protect the human rights... of its citizens."[22] And rights can be enforced.

Various climate lawsuits have now been brought, most in response to the lack of legal regulation for emissions and approaches to the consequences of climate change. The offending gaps in legislation violate the rights of future generations or ignore the needs of individual groups, preventing human-rights obligations from being met. One prominent example is Luisa Neubauer's successful lawsuit against the German federal government. The judgment handed down by the Federal Constitutional Court on April 29, 2021, called on the government to revise climate laws and expand them to include the rights of future generations.

At the time of writing, another suit is pending with the European Court of Human Rights (ECHR). This was submitted by an association of senior women from Switzerland, who argue that older women suffer particularly from the consequences of extreme heat and that the measures enacted by the Swiss government to reduce emissions are not enough to moderate future heat waves. The Swiss courts may have rejected the suit, but it has been taken on by the ECHR. Whatever the eventual verdict, this in itself is a success, as it shows a change in mind-set among judges and lawyers in particular and civil society in general, conceding that climate protection plays an important role in realizing human rights.

Other forms of climate lawsuits are mainly aimed at securing compensation or direct financial support for local measures to adapt to the consequences of climate change. One such case being negotiated in Germany is drawing plenty of attention. Saúl Luciano Lliuya, a Peruvian farmer, is demanding that energy company RWE pay 17,000 euros (in line with its contribution to global greenhouse gas emissions) toward a dam that would protect his town from flooding from a glacier lake.[23] If the Higher Regional Court Hamm finds RWE guilty, it will be the first time that the business model of extracting fossil fuels for burning is declared problematic in a legal capacity. The actual damage in this case is tiny (17,000 euros is nothing for a company like RWE) but the strategic consequences of this lawsuit are enormous. Companies like RWE would then no longer be able to shirk their liability for endangering human lives, as the door would open for countless citizens across the world to claim compensation for the damage they have incurred. Lliuya's case would set a precedent and mean that laws would have to be improved and applied more consistently. This could also reduce the legal gap between international agreements and the obligations of individual states or companies; ultimately, the case against RWE is about getting a company based in Germany to align its global business model with the objectives of the Paris Agreement and to no longer rely on burning fossil fuels. Lliuya's case is one of the first to cross national borders, as it targets one of the corporations that bears significant responsibility for climate change due to its business model and, above all, the large-scale lobbying in which it continues to engage. Almost all other lawsuits of this type have been and are being pursued in the U.S. At the time of writing, thirty-five out of more than a thousand climate lawsuits on specific damage are pending—for example, the suit brought by an Oregon county against companies such as Chevron and ExxonMobil for their contribution to the catastrophic 2021 heat wave. Like Luisa Neubauer's case, most lawsuits outside the U.S. (currently almost a thousand worldwide, less than in the U.S. alone) are aimed at national governments. But there are a few suits that have been brought before international courts, such as those against Bolsonaro.[24]

LAWSUITS FILED
AGAINST BOLSONARO

In robust democracies, lawsuits are a tried-and-tested method of achieving progress in the fight against climate change. But what is it like in autocratically inclined countries, where politicians not only let legal negligence slide but actually tolerate criminal actions against their own people, as was the case under Bolsonaro's rule? Will a government that doesn't feel obligated to observe its own laws even listen to the Supreme Court?

On June 5, 2020, after Bolsonaro had been in office for a year and a half, the Brazilian Socialist Party and six other parties submitted an injunction action to Brazil's Supreme Federal Court under the auspices of the Sustainability Network (Rede Sustentabilidade).[25] The complainants argued that the National

Climate Fund Project (Fundo Clima) set up in 2009 as part of Brazil's national climate plan was suspended in 2019; consequently, no more annual plans had been prepared and the corresponding money for projects to mitigate climate change and for appropriate adaptation measures had been held back.

The lawsuit invokes the duty of the government and local authorities to "protect the environment and fight pollution in any of its forms" and to "preserve forests, fauna and flora." It also references the principle of precaution enshrined in Brazil's constitution. Article 225 regulates the duties of the state, which include preserving and restoring natural ecosystems and identifying spaces that are to receive special protection. The aim of the lawsuit was to obtain a preliminary injunction to force the state to reactivate the Climate Fund and to provide it with appropriate resources to finance reforestation projects.

The suit was successful. On July 1, 2022 (while Bolsonaro was still in office), Brazil's Supreme Federal Court became the first in the world to explicitly recognize the Paris Agreement as a human-rights treaty, laying the legal foundation to challenge failures in the fight against climate change that endanger human lives as a result. Astonishingly, this verdict—with implications that cannot be overestimated—barely made headlines in the Global North, another sign that climate change is perceived differently here than in the Global South. For the Global North, aiming to limit global warming to 2°C (3.6°F) by 2100 seems more like an economic cost-benefit assessment. The damages anticipated in such assessments won't exceed the profits from burning fossil fuels until global warming passes 2°C. But like reports drawn up by major insurance companies, these assessments are based on physical damage to insured infrastructure—houses, public roads, power lines, and water pipes. They don't count human lives, ecosystems, or any damage to health, livelihoods, and culture. The same approach is used in political discussions on the implementation of the Paris Agreement, which focus mainly on physical parameters like sea level height, the intensity of a drought, or the amount of water in once-in-a-century rainfall—primarily to the detriment of those who are already suffering the consequences of climate change.

The verdict of the Brazilian judges is also pioneering because they did something that every court in the world could do whenever they lack knowledge of the matter being adjudicated. Beginning in September 2020, the Supreme Federal Court held public hearings to which they invited scientists and representatives of civil society and Indigenous groups. This provided a far more detailed insight than the usual expert reports and questioning of witnesses. Once the court had informed itself thoroughly of the underlying problems and available evidence in the lawsuit, it called on the government and the National Bank for Economic and Social Development (BNDES) to provide comprehensive information on the money allocated to the Climate Fund and the total amount spent on climate-related projects.

Only once all this information had been weighed up did ten of the eleven judges vote in favor of the injunction. This demonstrates the incredible importance and power of an independent judiciary. In his reasoning, presiding judge Luís Roberto Barroso observed the enormous increase in deforestation in the Brazilian Amazon region in 2021—a problem that, as we have seen, continues despite the change in government. Brazil is the world's fifth-largest emitter of CO_2, and deforestation is its greatest source of emissions. The Brazilian government has been asked to completely reactivate its Climate Fund. As the verdict states, "treaties on environmental law are a species of the genus human rights treaties and enjoy, for this reason, supranational status. Thus, there is no legally valid option of simply omitting to combat climate change."[26]

This has significant implications for national and international law; although the basic idea of the Paris Agreement is to protect people and their rights, this is only explicitly mentioned once in the agreement. The preamble states that the fight against climate change is a common concern of humankind and that the parties to the agreement should consider their respective obligations on human rights when taking appropriate measures.[27] In other words, those who do not adhere to the Paris Agreement will be guilty of blatant human rights violations. In view of this somewhat reserved phrasing, a verdict like that from Brazil is extremely important and could prove to be a crucial step on the road to placing justice at the center of climate policy worldwide. But the road remains rocky; even as the lawsuit was under way, the forest was being logged at ever-increasing speed. When the rainforest protection laws enacted before Bolsonaro's term were repealed, illegal logging suddenly became legal, and the new government under Lula da Silva is finding it difficult to undo these arbitrary measures.[28] But there is cause for hope. Lula has declared saving the rainforest to be a key objective of his time in office. And further climate lawsuits are pending with the Brazilian Supreme Federal Court, so there may soon be more instruments available for establishing climate justice.

INTERNATIONAL COURTS

Lawsuits are usually brought before Brazil's Supreme Federal Court (a constitutional court) by federal governments and authorities. Individuals can too, but they have to prove that their applicable constitutional rights in Brazil are being violated, which isn't always easy. Unlike national courts—like the Supreme Court—the International Criminal Court (ICC) in The Hague has a whole other remit and jurisdiction determined by the Rome Statute, a treaty signed by many nations. The first article states that the ICC is a permanent institution with the authority to pass judgment on the most serious crimes of international concern.[29] The ICC focuses on genocide, war crimes, and crimes against humanity. Individuals cannot file suits here, but one of the roles of the ICC's Office of the Prosecutor (OTP) is to ensure their concerns can be heard. The OTP conducts investigations on people suspected of crimes that can be prosecuted under international criminal law and brings charges in cases of justified suspicion.

On October 12, 2021, the OTP received a communication from a specially founded NGO calling for an investigation into the Brazilian president's role in the persistent deforestation of the Amazon rainforest and associated crimes against humanity. Specifically, it stated that President Bolsonaro had promoted a "widespread attack" on the Amazon and "those who defend and depend upon it," in "a clear and extant threat to humanity itself."[30] This wasn't just about the human rights of the Brazilian people, but of all humanity, which would be a serious crime under international law. The complaint argues that global climate security depends on the Amazon because it plays a key role in regulating global temperatures and weather systems and that "severe damage... stemming from mass deforestation, conversion of deforested land to cattle ranching, and vast, intentional forest fires" has significantly disrupted the functioning of this critical ecosystem and gradually turned the Amazon from a carbon sink to a carbon source. This argument is founded on the additional 180,000 (at least) predicted heat-related deaths by 2100 already mentioned and the effects of forest fires that can be demonstrably attributed to climate change. The communication also stated that the Brazilian authorities were unable to enforce international laws against Bolsonaro and his allies and that it was therefore necessary for the ICC to intervene because human rights were being violated worldwide. It remains to be seen whether the ICC will accept this argument. The appeal to the International Criminal Court is undoubtedly important; at the very least, it raises awareness of the grievances and may open up new judicial territory—up to now, the only successful court cases have aimed for more ambitious reductions in emissions. The courts have so far rejected claims for compensation for losses due to climate change. A positive sign from the ICC would encourage judges in general and would be an important step toward greater—perhaps significantly greater—justice.

Why do judges often lack the courage to take on such cases or rule in their favor? Do they genuinely lack the courage, or are there fundamental legislative obstacles? I tried to tackle this question in a study conducted with colleagues from law faculties at various universities.[31]

In this study, we identified two central themes or causes that are currently hindering the judicial handling of damage caused by climate change, thus preventing real progress in matters of climate justice. Borrowing from the American Latvian philosopher Judith Shklar, we named one of these motives the "social superstructure narrative," by which we mean social perception of climate change—what most nonexperts (including judges and other law practitioners) understand climate change to be. For them, climate change is apparently less a process of irreversible environmental destruction and current injustice, and more a future problem that in principle we can largely avoid by reducing emissions or adapting. The second aspect of our study relates to causality—that is, causes and effects in the context of climate change. The general public may be aware that intensive research is being conducted into the causes of the climate crisis, but the results of this research are often presented in a way most people can't understand—not least the results of very new research on the specific losses and damages caused by climate change.

Social perception is important for every climate lawsuit because in every case, the judge must first decide whether the matter in dispute actually belongs in the courtroom. This depends to a large extent on generally recognized norms and ideas of morality; the political, social, and cultural context plays a central role in which judgments are passed and whether lawsuits are taken on in the first place. Law is both an idealized notion of order and a cultural and social practice. Take the following example: in a legendary British case from 1956, a plaintiff won his argument that the sight of sex workers and their clients impaired his ability to use his home (Thompson-Schwab v. Costaki). The judge followed his reasoning, and the defendants were held accountable for nuisance because the mere existence of sex workers was regarded as indecent. The verdict would certainly be different today.

Examples like this underline that the law is much more than its wording. The normative context is crucial; in other words, legislation and jurisprudence never take place in a vacuum. Judges pass verdicts in a specific political and social environment, and the success of a legal dispute without influential precedents depends not least on the values of the judges. With a few exceptions, judges have completely internalized the colonial-fossil narrative and see no legal problem whatsoever with climate change being intensified through the development of new oil fields, or inner-city air being contaminated by increasing traffic. But this isn't set in stone. It can be changed by science, the media, activists, or by disasters; so can the "social superstructure narrative"—and with it, the interpretation of legal texts. Developments in Australia in recent years show how a country generally known for having climate deniers in high political offices can place climate justice at the center of society, thanks to a mix of angry weather and robust courts with independent judges.

7

FROM PAWN TO GAME CHANGER

Australia

"We've got people who have been in the brigade for twenty-five years who said they've never seen anything like that; it traveled four kilometers [two and a half miles] in about twenty minutes."

"When I see my colleagues from my brigade jump into a plane or a bus... to go into battle against an unprecedented enemy of catastrophic proportions, I wonder if we might ever see them again. And they are my friends and wonderful people."

"When my RFS pager goes off in the middle of a hot, blustery, severe-fire-danger day and I have to rush off to a bushfire, and as I am sitting in the truck proceeding under sirens and lights to the fire, I wonder if this might be my last day too."[1]

"This is climate change in its most fundamental form. This is right in our face. We're living it."[2]

AUSTRALIA IS LARGELY made up of desert and semidesert, which aren't exactly fit for human habitation. The south-western and southeastern tips are the exceptions, and so this is where most of the people, animals, and plants reside. This is also where, in the summer of 2019–20, raging bushfires reached unprecedented levels. Bushfires are already part of Australia's inhospitable climate, which is why it appears to be disproportionately susceptible to the impact of climate change. Its weather is always at the limits of what humans can tolerate, so even the smallest changes can have devastating effects. The argument frequently used by climate-change skeptics—that extreme weather has always existed and therefore isn't evidence of climate change—sounds like a cruel joke in Australia.

But even in 2020, as the horrific fires dominated the head-lines, more Australians denied the existence of human-caused climate change than in practically any other country.[3] All the same, this still meant that, as a representative survey found, 58 percent of Australians considered climate change a serious problem, with opinions varying widely depending on political orientation. People to the left of the political spectrum were more concerned about climate change and found reporting to be more accurate and informative than right-of-center Austra-lians, who not only appeared less concerned, but also thought the media's depiction of climate change was either wrong or at least exaggerated—saying that too much attention is given to the topic. The survey identified a similar divide between the generations; older people were less concerned and had a more negative attitude to climate change reporting than younger people.

None of this is particularly surprising.

As in other countries with pronounced climate skepticism, Australian climate-change deniers consume news sources that align with their beliefs. Around a third of those surveyed used commercial media like Sky News and Fox News and believed that climate change isn't relevant. These two broadcasters in particular, which are part of News Corp (majority shareholder: Rupert Murdoch), are known for provocative presenters who call into question the scientific consensus on climate change and place little value on facts overall. One example that made international headlines for all the wrong reasons was a 2016 debate arranged by the Australian Broadcasting Corporation with the popular British physicist Brian Cox to discuss whether the time had come to take climate change seriously. The program discussed well-known false statements from climate deniers and ignored any real questions on how to deal with climate change. Once again, this program suggested to the public that there are no facts, only opinions.[4]

The survey also revealed a surprising difference between rural and urban news consumers. Apparently, people in the countryside were less worried about climate change than city dwellers. What makes this so astonishing is that when the survey was conducted, practically the entire densely populated section of rural Australia was going up in flames—you'd assume that the people most affected by the fires would feel more threatened by climate change.

One possible explanation for this phenomenon is that the rural population of Australia, as in other industrialized countries, is older and more conservative than the rest of the country. But this alone doesn't explain why climate change was denied for so long and so stubbornly in Australia (and sometimes still is). There are two main reasons that, while not exclusive to Australia, have really taken hold there: the influential fossil energy lobby and a media landscape in which opinions aren't as wide-ranging as its plethora of publications might suggest.

COAL AND GAS INDUSTRY

In 2021, Australia was the world's largest exporter of coal and the second-largest exporter of gas.[5] The continent's economic system has been shaped by colonialism and, like many other capitalist democracies, is built on fossil energy. In the Australian public consciousness (as elsewhere), this has long been synonymous with a high standard of living. Because of this, the populace doesn't question the constant expansion of the fossil economy, seeing it as a legitimate regime anchored in specific habits such as driving cars, flying, and air conditioning.

Australia's fossil industry has harnessed this inertia with particular success, its strategy one of perfidy: although Australian business tycoons publicly recognized climate change earlier than industry representatives in other industrialized countries, they didn't do anything about it. In 2008, launching a PR campaign against the then Labor government's proposals on CO_2 pricing, the Australian Coal Association (ACA) hastily conceded that "climate change is a big problem" and that "science shows human activities are contributing to it through... burning fossil fuels like oil, natural gas, and coal."[6] The industry promised to advocate for climate protection measures, but behind closed doors it lobbied for a different outcome, trying to keep emission reduction obligations as low and vague as possible and to hinder the expansion of renewable energies. In public, lip service to climate change was consistently contradicted by statements declaring the nation's main concerns to be energy security, low electricity costs, and maintaining revenue from coal exports—all of which were apparently only possible with fossil fuels. Ralph Hillman, head of the Australian Coal Association, took every opportunity to emphasize that "fossil fuels underpin Australia's economic growth and will do so for the foreseeable future."[7]

The coal industry had another strategy to undermine political efforts for greater climate protection, and that was to demand a "measured response" to calls for reduced emissions. It argued in favor of a state that generally holds its punches and doesn't need to govern all that much because the energy industry will act rationally and sensibly of its own accord. What the coal industry failed to mention was that, in essence, it interpreted "sensible" to mean maximizing profits, not the interests of the people. From this perspective, the development of new coal mines, gas fields, and offshore oil drilling is "sensible" because economic growth has been portrayed as possible only by continuing the same traditional industries unchanged, and the fossil industry pretends that burning carbon is the only way to satisfy the rising demand for energy. Unfortunately, some of the people who actually want to advocate for climate protection have gotten caught up in this persistent narrative.

Industry rhetoric continues to claim that the economy has the theoretical potential to innovate and stave off climate change with technological solutions. So we don't have to do anything right now; we can simply carry on as normal. It's best to let the industry grow unchecked because we'll be able to reduce emissions when the innovations eventually kick in. But for now, there must be as little regulation as possible; otherwise there won't be enough capital for research and development.

The Australian fossil-fuel industry put up another smokescreen by talking about "clean coal." Specifically, it claimed that Australian coal is "cleaner" than its global rivals because it offers a particularly high energy density and thus produces 10–20 percent fewer CO_2 emissions per ton burned. And so, on balance, supporting coal in Australia would actually be good for the climate. Obviously, this is utter nonsense. Coal from Australia is the opposite of clean, just like all other coal, both in terms of climate change and all other pollutants. The concept of "clean coal" was invented by R&R Partners, an American PR firm whose campaigns had already helped the tobacco industry to stop smoking bans from being introduced.[8] The concept spread particularly well in Australia thanks to the Murdoch media, but it remains popular in the U.S. too after Donald Trump obligingly embraced the rhetoric in his campaigns. A similar strategy was applied to justify support for gas and oil and tout this as the sensible choice; apparently, these energy sources are less harmful to the climate than coal and will replace it in the future.

This argument fell on fertile ground in Australia of all places, with its devastating forest fires, despite the fact that prosperity doesn't protect against the consequences of climate change as efficiently here as it does in most other highly industrialized countries. You might be able to evade the heat in well-insulated houses, but there's no effective method against the toxic smoke that indirectly causes many deaths; breathing in the smoke, often for weeks, triggers respiratory diseases, and the fine particles can lead to coronary shocks such as heart attacks and strokes (see chapter 6). Increases in forest and bush fires have led to a measurable increase in both hospital admissions (42 percent) and deaths (up to 92 percent).[9]

Nevertheless, the ideology of the fossil-fuel industry influences the most endangered groups—older people and the rural population who live near the bush. Australia's media landscape is certainly one reason for this, as it has vigorously supported the strategy of the fossil-fuel industry. Radio and television broadcasters disseminate the lobby's spurious arguments and also provide a major platform for climate denial. This impedes scientific journalism in general and climate reporting in particular, steering the focus away from the actual problems. Relaying scientific findings to the public is an art in itself, treading a fine line between complexity and comprehension. But doing all this while competing with messages that portray climate change as a conspiracy becomes a Sisyphean task. Again and again, fake news must be exposed as such—and as soon as it is, the next fake report has already been published. It has been claimed (not just in Australia) that back in the 1970s, scientists believed we were about to enter a new ice age.[10] Obviously this wasn't true, but journalists had to expend a lot of time and effort doing the research to expose the lie. That time could have been spent on substantive climate change reporting, and despite being corrected, the misinformation still causes confusion when it resurfaces, slightly altered and with vague references to hacked emails and withheld data. Further, misinformation also spreads quickly, such as the claim that scientists are simply ignoring many factors—like sunspots and volcanic activity—in their climate projections. This isn't true either, but again, reputable journalists were compelled to conduct extensive research to identify the sources of the false claims and set the record straight, research that had nothing to do with the pressing problems of climate change. Gradually, the multitude of false reports in Australian media and on internet platforms have created a distorted picture of climate change, ultimately meaning that even many of those affected continued to deny the existence of climate change during one of the country's worst bushfire disasters.

What does this have to do with climate justice, or rather injustice?

A great deal! As the media and fossil-fuel industry stood shoulder to shoulder, casting doubt on the reality of climate change, Australian politics were biased toward the interests of industry. The fossil narrative gradually took hold not only in the minds of politicians and their voters, but also in decision-makers at various levels of society.

FIRE

Until 2023, 2019 was the hottest year in Australia at 1.52°C (2.74°F) above preindustrial levels and, with 40 percent less rainfall than the long-standing average, also its driest year since 1900 (when standardized temperature and rainfall measurements began).[11] The years 2017 and 2018 were also extremely dry throughout vast swaths of the country, creating ideal conditions for the bushfires that began in November 2019 (earlier than usual) and quickly spread to densely populated areas. At least thirty-four people died in the flames themselves, and while the victims of the deadly air quality caused by weeks of strong smoke pollution haven't been counted, the number is certain to be many times higher. Almost six thousand buildings were destroyed by the fires. It is estimated that between 500 million and 1.5 billion wild animals died, along with tens of thousands of farm animals.[12] The bushfires also had a significant economic impact (including enormous insurance payouts) and long-lasting effects on the health—both physical and mental—of all those exposed to the smoke. Trauma, anxiety disorders, and PTSD were triggered by the smoke, the fire, and the powerlessness that people felt in the face of the encroaching flames.

The high speed at which the bushfires spread often made it even more difficult than normal for the fire service to fight the fires and save lives. Energy, fuel, and food supplies were disrupted on a daily basis. The need to close roads cut some communities off completely. The firefighters quoted at the start of this chapter bore witness to their feelings of powerlessness and fear for their own lives and those of their colleagues. At the end of this chapter, we will see how they turned this powerlessness into power.

The development and scale of fires are determined by many factors, including the quantity and composition of combustibles, temperatures, oxygen, rainfall, humidity, wind direction and speed, and the specific landscape dynamics. Australians have a wealth of experience in how best to stop bushfires. Aboriginal Australians have been practicing fire management techniques for millennia. But as climate change has made its presence increasingly felt in the last seventy years, the conditions for fires have changed. Many parts of Australia have become hotter and drier, meaning that bushfires occur more often and are more intense. They release significant amounts of greenhouse gases, further strengthening global warming. In response, some scientists have declared that the "Pyrocene" age is upon us—a new era of increased fire risk.[13] My team and I were able to confirm this with a study in which we established that fires such as the "Black Summer" of 2019–20 had become more than 30 percent more likely due to climate change.[14]

MEDIA

Fierce debates raged among the Australian public on the causes of and associated blame for the bushfires of 2019–20. The link between the fires and climate change became a political issue.[15] In an interview with the Australian Broadcasting Corporation (ABC), then–Deputy Prime Minister Michael McCormack said it was "disgraceful" and "disgusting" that reports were linking the bushfires with climate change. According to him, people needed help, not "the ravings of some pure enlightened and woke capital city greenies." Gladys Berejiklian, premier of New South Wales, also said it was "inappropriate" and "disappointing" to talk about climate change while the fires were still raging. Carol Sparks, mayor of Glen Innes, a town in rural New South

Wales devastated by the bushfires, emphatically contradicted their statements, telling ABC that "Michael McCormack needs to read the science." She stated that climate change was "not a political thing" but "scientific fact."[16]

The more intense the fires became, the more things heated up in the media too. Reports were split along ideological lines. The public service broadcasters ABC and SBS (Special Broadcasting Service) and the newspapers *The Age*, the *Sydney Morning Herald*, and *Guardian Australia* portrayed the fires overwhelmingly as the result of climate change. *The Guardian* declared that "despite the political smokescreen," the link between the fires and global warming was "unequivocal." ABC also characterized the fires as "a glimpse of the horrors of climate change's crescendoing impact."[17] The news agencies owned by Rupert Murdoch's News Corp Australia took a very different approach and tried to downplay the fires. When the scale of the disaster became apparent on the last day of 2019, Murdoch's *The Australian* published an article comparing the fires with earlier incidents and claiming that such conditions were "not unprecedented" and the fires were "nothing new." News Corp's Sky News repeatedly used the term "climate alarmists" whenever climate change was discussed.[18]

The longer the fires went on, the more Murdoch's section of the media landscape became determined not to mention climate change at all in connection with the fires and to defame anyone who did. The conservative media insisted that the fires were "normal" for Australia and ascribed their severity to failures in fire prevention. For example, they said that targeted felling had been prevented, for which they blamed "Greens policies."[19] Accordingly, they claimed that climate protection and emissions reductions were superfluous. Minister for Energy Angus Taylor took the same line, publicly declaring that with "only 1.3 percent of global emissions," Australia could not "have a meaningful impact"[20] on global warming and thus did not need to reduce emissions.

For most international media, there was no doubt that the fires in Australia were stoked by global warming and further intensified by the climate change denials of the Australian government (then led by Prime Minister Scott Morrison). *Deutsche Welle* called Australia a "notorious climate offender," and the *New York Times* declared the country to be "committing climate suicide" with its misguided climate policies and continued coal mining.[21] Other Western media reminded their audiences that two years before the devastating fires, Morrison had proudly brought a fist-sized piece of Australian coal into the parliament. In the U.K., *The Independent* published an article entitled "This is what a climate crisis looks like."[22]

Naturally, misinformation about the fires also spread on social media. The hashtag #ArsonEmergency went viral, telling people that the fires were largely caused by arson.[23] Although this hashtag was quickly revealed to be the work of bots, its effect on public reporting was barely diminished. Many of Murdoch's media outlets took to the arson idea with gusto, some commentators embellishing it with hints that Australia burning would certainly be handy for the "climate alarmists," as it would help them spread panic much more effectively. By sowing seeds of doubt on the link between climate change and the bushfires, the conservative media managed to skew the public discourse. The bushfires weren't a wake-up call to those sections of the population that tended toward climate skepticism; instead, the fires were exploited to bolster their misgivings.

In Germany too, the tactic of drawing attention to arson is used to obscure the connection with climate change. Jörg Kachelmann, a former weather forecaster with the channel ARD, spread a similar story relating to fires in Canada in 2023. This was in response to a study that my team and I had published in

August 2023, showing that climate change had made the fire weather conditions 50 percent more intense. Naturally, this has nothing to do with whether and how a fire is lit. In Canada, fires are often started by lightning strikes, and sometimes by arsonists. But the problem is that there are always sources of ignition, and the resulting fires are more dramatic due to climate change—in Canada, in Australia, and in Europe. So unfortunately, this diversionary tactic isn't limited to Australia.[24]

Luckily, an increasing number of people in Australia are now taking climate change seriously and getting their information from the international press and the Australian Bureau of Meteorology, which employs a group of scientists focused on the link between weather extremes and climate change. They regularly publish studies showing how much record heat and forest-fire risks have increased in recent years, and they provide clear evidence of climate change's role in this process. Although the Bureau of Meteorology is a state institution, its employees have advisory roles at most and can only influence policy if the government desires it.

During the 2019–20 fire season, the government largely ignored warnings from experts and the protests of those Australians who campaign for greater climate protection—including many of the firefighters who have to face the growing strength of ever-larger bushfires. In many parts of Australia, volunteers too put their lives on the line and witnessed terrible things. The fires had significant psychological repercussions for those who lost their homes, and sometimes more, and also affected those who wanted to help but couldn't. It may take years to process the images of destruction and suffering, and this may be accompanied by post-traumatic stress disorders and depression.

Research largely agrees that personal experiences tend to confirm, rather than negate, our world views.[25] So those who deny the existence of climate change tend not to be swayed by disasters. But perilous situations like bushfires engender a greater sense of urgency in people who don't doubt the existence of human-caused climate change. Their conviction that climate change has arrived—that it is happening here and now—grows stronger. For them, climate change is no longer just a theoretical concept; it becomes a lived experience that unleashes their desire to make a difference. The resulting climate activism can take many forms, from public protests to civil disobedience, from political engagement to climate lawsuits. The latter has been extremely successful in Australia and probably contributed to the election of a new government on May 21, 2022, which will hopefully make a better contribution to global climate justice. So far, there has been nothing to warrant this hope, but it does mean that climate change has evolved from political pawn to political game changer.

COURTS

Survivors of the 2019–20 bushfires filed the first successful lawsuit that explicitly invokes scientific studies quantifying the role of climate change in a specific extreme event.[26] The plaintiffs (bushfire survivors represented by the New South Wales Environmental Defender's Office, an NGO specializing in legal help for environmental matters) argued that climate change made the bushfires more likely and more intense. The case was tried from April 2020 to August 2021, as the plaintiffs called on the Environmental Protection Authority (EPA) to issue binding guidelines that can help to reduce greenhouse gas emissions to net zero by 2050. The EPA may have supported the Paris Agreement, but in the view of the plaintiffs, it had not yet developed any plans for national implementation.

The EPA disagreed, but the court still decided to try the case. Penny Sackett, an Australian climate scientist from the

Australian National University in Canberra, served as a witness for the plaintiffs. She was invited to talk about the status of climate change in Australia and to answer the question of whether emissions in the region are on track to limit global warming to 1.5°C (2.7°F). According to the plaintiffs, it was a novelty for evidence for climate change to be admitted before an Australian court to prove that the government had not fulfilled its legal obligation.

The presiding judge, Justice Brian Preston, followed the plaintiffs' reasoning in his verdict, determining that climate change presents a real and significant threat and that the EPA has not fulfilled its duty to protect. According to Preston's judgment, none of the measures it has mandated offer sufficient protection against climate change and its consequences. Previous EPA guidelines were "directed towards ancillary or insignificant causes or consequences of climate change."[27] So far, Anthony Albanese's government has not contested the ruling and, according to media reports, does not intend to do so; instead, it promises to do everything necessary to see it through.

Another interesting aspect of the verdict was that Justice Preston used it as an opportunity to instruct future plaintiffs on how to file more promising lawsuits. In the second paragraph of his grounds for judgment, he writes that it is difficult to allow actions that are based not on facts, but on contestable opinions, regardless of how justified their core concerns. As an example, he quotes the following statement from the notice to admit facts: "Unregulated release of greenhouse gases is the greatest threat to the environment of NSW."[28] While nobody is seriously doubting that greenhouse gas emissions are a threat to the environment, as soon as the plaintiffs declared this to be the "greatest" threat, they jeopardized their chance of success. It will be difficult to prove that greenhouse gas emissions are a greater danger than any other.

Preston was sympathetic to the lawsuit because he is one of the few judges in the world who has engaged with climate change in detail and therefore knows where scientific evidence is strong or thin, and for which statements. He is an exception to the rule; most judges, lawyers, and even activists know no more about climate change than what the media reports. How do I know? In one study, my colleagues and I looked at more than eighty climate lawsuits and found that not a single one had adequately described the state of research.[29] Errors were often made similar to those criticized by Justice Preston, or claims were made about the link between climate change and specific damage without being able to provide relevant evidence—for example, because they chose a region for which no data is available. Finally, climate lawsuits repeatedly cited studies that used methods that have since become obsolete, decades ago in some cases.

In his decision, Justice Preston used the analogy of a tennis match to illustrate what had gone wrong. He compared the length and intensity of rallies to exchanges between lawyers facing each other in court, taking turns to provide evidence and respond accordingly. He went on to say that the legal dispute with the EPA—with statements from the bushfire survivors and evasive answers from the EPA—had more in common with golf than tennis; the two parties were playing on totally different parts of the course, and so there was no actual contest between them. Justice Preston sidestepped the dilemma of superficial knowledge of climate matters by inviting proven experts to the proceedings against the EPA, and distilled the content of the complaint into five precise questions:

1. To what extent have anthropogenic greenhouse gas emissions caused a significant increase in extreme fires and extended the fire season across Australia?

2. Can the world achieve a global rise in temperature of no more than 1.5°C (2.7°F) above preindustrial levels? If yes, how; and if not, why not?

3. Do the current regulatory conditions for emissions in Australia align with the limitation of the global rise in temperature to no more than 1.5°C?

4. Does the current reduction in emissions for New South Wales align with the limitation of the global rise in temperature to no more than 1.5°C?

5. Are the targets and guidelines required by the respondent suitable for:

   a. reducing direct and indirect sources of greenhouse gas emissions in a manner consistent with the limitation of the global rise in temperature to 1.5°C compared with preindustrial levels?

   b. mitigating against the threat posed by climate change to the environment and the people of New South Wales? [30]

The experts' responses to these points supported the charge. The EPA was given the opportunity to respond to the accusations, but its representatives were unable to put forward any persuasive counterarguments, and so the climate lawsuit was ultimately sustained.

Justice Preston also presided over another case relating to climate issues. The mining giant Gloucester Resources Limited sued the Australian minister for planning for turning down its application to build an opencut coal mine in New South Wales. The Rocky Hill coal project was supposed to produce 21 million tons of coal over sixteen years. After weighing up the costs and benefits of the project, including the climate change impact to be expected due to the direct and indirect greenhouse gas emissions from the mines, Justice Preston ruled that the project was not in the public interest.

The company appealed and argued that Australian coal is "cleaner" than other products available on the global market. The court maintained its position, and the minister again confirmed that the application was declined. Explaining his verdict, Justice Preston stated that

> the negative impacts of the Project, including the planning impacts on the existing, approved and likely preferred land uses, the visual impacts, the amenity impacts of noise and dust that cause social impacts, other social impacts, and climate change impacts, outweigh the economic and other public benefits of the Project.[31]

In these two verdicts, Justice Preston shares the view of the scientific community, but unfortunately he is way ahead of his time. Plaintiffs may lack knowledge of the link between extreme weather, the resulting damage, and climate change (as we have seen in notices to admit facts), but so do judges. And they are the ones who ultimately decide whether a lawsuit is heard. As we saw in the previous chapter, their judgments depend on generally accepted norms. The political, social, and cultural context that determines the wording and interpretation of a law is enormously important in the development of legal practice.

The superstructure narrative is extremely sluggish in a society that remains steeped in the tradition of fossil-fuel dependency and is kept there by lobbies and the media. But climate lawsuits might get it moving. They have triggered political and legal changes in Australia and other countries that have seen legal triumphs. But we are nowhere near the radical societal changes required for a fair approach to the consequences of climate change. Neither the law nor science can bring about these changes alone, but they do play a crucial role. However, corporations and governments are not at a particularly high risk of being sued right now. And so they cling to business models that continue fossil-fuel dependency. In most companies, it is the communications department that figures out how to deal with climate change, not the CEOs who determine the business model.

While successful lawsuits have heightened awareness of injustice in matters of climate change, they have so far had little impact at a macrosocial level. This is also clear from the fact that, so far, legal claims for damages currently being caused by climate change have been few and largely unsuccessful. And there are no binding control mechanisms for climate risks in economic contexts. In the Global North in particular, the public continues to view climate change as a future problem.

This is due both to the active efforts of lobby organizations and to the lack of broad access to education on climate change as it relates to both science and law. Climate change is often included in educational concepts in condensed form only, and even climate activists and NGO workers can't usually tell the difference between scientific facts and speculation.

In many instances, the climate science community is definitely in a position to prove the specific and future damage that is and will be caused by particular emitters. This information should be presented in a comprehensible and engaging way and form an integral part of climate change education in schools, universities, the media, and courtrooms. In London, for example, we have launched an initiative to provide scientific input to a network of legal practitioners. Germany has similar projects providing basic scientific concepts for legal proceedings.[32]

Hopefully projects like these will have a snowball effect if court cases that successfully harness climate science then help to change the superstructure narrative in which climate change is assessed. This sort of change will then improve the chances of further legal successes. And more successes are urgently required, given that climate change is already having a huge impact, particularly on those groups in society that are already struggling.

Of course, we can all help to raise social awareness of the consequences of climate change by learning about the latest research and passing on this knowledge. This includes people in creative fields, artists, and filmmakers, who can have a major influence on public perception. Why did *Don't Look Up*, a hit Netflix movie, make do with comparing climate change to an asteroid, an analogy that doesn't help at all? Why doesn't Hollywood produce a thrilling blockbuster about a lawsuit against ExxonMobil? Climate change isn't an undeserved stroke of misfortune; above all else, it is injustice. It is fundamental that we understand this. And to bring charges against this injustice, we need to realize that the entire chain of causality from emissions to specific damage can often be precisely documented. But not every effect of an extreme weather or climate-related event can be attributed to human-caused climate change. And there are cases in which a clear link cannot be made.

This knowledge is comparatively new and therefore not yet commonly known. There is also currently no comprehensive inventory of the effects of climate change, only individual case studies such as that on the damage caused by Hurricane Sandy in New York in 2012—an event clearly attributable to human-caused climate change. The proportion of the damage from human-induced climate change amounted to 8 billion U.S. dollars.[33] These days, such things can be calculated.

Nevertheless, various interest groups—most notably the lobbies for the coal, oil, and car industries—continue to cast doubt on the reliability of climate studies and their findings. Some people claim that these are only models from which no accurate results can be derived. But the data is less prone to errors than other scientific disciplines, and climate research can provide concrete findings on many specific questions. Nevertheless, there are numerous questions that can't yet be answered. This isn't a problem particular to climate research, but the way to progress found in all of science—one answer tends to throw up ten new questions.

Naturally, probabilities play a major role in climate research too, but this isn't a flaw, just good normal scientific practice. The entire discipline of quantum physics is based on probabilities. And courts accept probabilities in principle—for example, when it comes to health damage. This means nobody has to prove to a court that it would be impossible to develop cancer without coming into contact with a carcinogenic substance. A similar approach is taken to the link between asbestos and lung disease: if a single employer isn't seen as the cause of the damage, then several employers can be liable, even if one is responsible for such a small proportion of the contact between the victim and the asbestos that, taken in isolation, it wouldn't have led to the disease.[34] The role of the courts is to interpret laws in such a way that they create a more just society.

However, neither the scientific nor the legal community alone can make climate change and its consequences the focus of general public attention. This epoch-making topic needs to be mobilized by large sections of the population. The developments in Australia since the 2019–20 bushfires give cause for hope. Demand for information on climate change has grown throughout Australian society, particularly during extreme weather events. As the bushfires raged, the proportion of heavy news users increased to 52 percent.[35] This is particularly worth mentioning because before the fires, Australians consumed less news than the rest of the world. It then climbed to sixth place in the global ranking of news consumption. And now that Australia has had two successful climate lawsuits and elected a government aware of the consequences of climate change, enthusiasm is waning for anti-science conspiracies. In 2022, 75 percent of the population saw climate change as a serious problem.[36] It's very possible that in the coming years, Australia will provide the world with less "clean coal" and more inspiration for climate justice.

PART IV

FLOOD

How Local Attitudes and Global Politics Are Saving and Destroying Livelihoods

8

GUILT AND RESPONSIBILITY

Germany

"The warning chain didn't work; one community didn't know how flooded the others were for a long time, no sirens went off, and the residents remained unaware for too long—this was a collective failure of politics and the authorities."[1]

"The children cried; they screamed in fear."[2]

"We are so thankful. But it will be a long time before we recover mentally, physically, and, above all, financially."[3]

"It could have been prevented… Her death was completely unnecessary."

"We only hope that everything is being done to ensure a disaster like this doesn't happen again."[4]

W HEN I WAS at school in Germany, more than once I heard my geography teacher say that our country was lucky to have no natural disasters. We had no tornadoes and only occasional, weak earthquakes, which caused minor property damage at most. And we didn't have to deal with volcanic eruptions either. This may all still be true, but as I know today, adverse weather phenomena can prove deadly at Central European latitudes too. The factors I explored in chapter 2 (on the northwestern U.S. and western Canada) apply to Germany and the United Kingdom as well. In the south of England, where I now live, more than three thousand people died in the heat of July 2022. But we don't see heat-related deaths, so for those who heard words of wisdom similar to those that I did at school, that sense of invulnerability felt by many Europeans won't be shaken soon by the hot summers of the twenty-first century.

But the summer of 2021 told a different story. The western German states of North Rhine-Westphalia and Rhineland-Palatinate (as well as parts of Luxembourg, Belgium, and the Netherlands) were given a dramatic lesson in the deadly potential of the weather in one of the world's richest countries, a place with an extremely humane climate. Between July 14 and 15, 2021, heavy rain caused by storm "Bernd" led to severe flooding and an alarming number of deaths.

It would be unfair to claim that my geography teacher lulled his students into a false—and deadly—sense of security. He didn't know any better, and he wasn't totally wrong, because what happened that summer may have been a disaster—but it wasn't a *natural* disaster.

THE FLOOD OF 2021

Storm "Bernd" approached Germany very slowly, giving it plenty of time to absorb warm and humid air in the Mediterranean before it rose to higher strata in the westerly low mountain ranges, cooling down as it did so. This culminated in widespread, persistent heavy rain, flooding small streams and creating flash floods. When the rain didn't ease, medium-sized rivers (such as the Ahr and the Erft) also burst their banks, and eventually the big rivers in the region (the Ruhr and the Meuse) did too, causing large-scale flooding in western Germany and Belgium. In the days after July 15, the high-pressure system "Dana" pushed "Bernd" toward Southeast Europe, leading to further heavy rain events in the east of Germany but with far less dramatic consequences.

There had been heavy rainfall even before this calamitous week, so in the affected areas of Germany and Belgium the ground was largely saturated, and it struggled to absorb any new rainwater. The ground was like a sodden sponge, particularly in Rhineland-Palatinate and southern Westphalia. The special geological features of the region around the Ahr River also played a role. The technical term is "Paleozoic siliciclastic sedimentary rock," which also contains limestone. It's important to know about this structure if we are to understand how the rain was able to trigger such devastating floods.

The ground on top of this rock has evolved to be relatively flat with low capacity for water storage. Topographically, the Ahr region is a plateau (generally 656–1,640 feet, or 200–500 meters) dominated by mountain ranges (up to 2,296 feet, or 700 meters) carved deep by a network of rivers during a phase of tectonic uplift. In many cases, this created funnel-shaped valleys with steep slopes. These V-shaped valleys are usually very narrow, meaning that the flowing water quickly swells and can overflow onto the banks. The steep slopes are also susceptible to falling debris, which causes additional damage when rocks or trees land on buildings, for example. Piles of debris can alter the flow of rivers and streams, and if things get really bad they can even block them, creating flood waves.

In contrast to the Ahr Valley, the Lower Rhine Bay (where the River Erft flows) has no naturally occurring steep slopes. But it does have slopes made by humans through mining, road and railroad embankments, and slag heaps.

These hydrogeological and topographical features make the region vulnerable to quickly rising water. The combination of extreme rain in July 2021 and already-saturated ground created the conditions for destructive flooding. This wasn't the first time the region had flooded; widespread flooding occurred after heavy summer rainfall in 1804, 1888, 1910, 2003, and 2016. But 2021 exceeded the historic high-water marks. Some flow-measuring stations were even carried away when rivers burst their banks at high speed. More than 180 people died, nine thousand houses were destroyed, and around seventeen thousand people lost all their possessions in the disaster.[5]

Ultimately, the amount of rain that falls in a region doesn't just depend on the type and speed of the storm; it is also strongly affected by regional characteristics like the height and conditions of the mountains. After the disaster, my team conducted an attribution study in collaboration with the German Weather Service (Deutscher Wetterdienst) in which we focused on two highly affected regions in the drainage area of the Ahr and Erft Rivers and of the Meuse, asking and answering the questions of whether and to what extent climate change has altered the intensity and frequency of an event like this. We also asked the same question for the entire area covered by the rain: To what extent does climate change alter the frequency and intensity of similar events in Western and Central Europe—that is, from north of the Alps up to the Netherlands? This area has similar rainfall characteristics to the regions affected by the floods in July 2021. We therefore looked at the whole region in which summer rain occurs due to the same meteorological processes. Further north, where it's colder, the transport of warm air to higher strata plays less of a role; further south, the Alps influence the properties of rainfall. Further to the east, the oceans are so far away that they have no influence at all.[6]

INFLUENCE OF CLIMATE CHANGE

Most of the rainfall responsible for the severe flooding in the Ahr and Erft region fell within a single day on July 14, 2021. It took two days in Belgium's Meuse region. Our study analyzed the quantities of rain in twenty-four and forty-eight hours respectively and revealed that climate change had played a considerable role. The relative frequency had increased by up to a factor of nine. So if the disastrous flooding were to occur statistically once every five hundred years in our current climatic conditions, it would have been nine times rarer without climate change and would only have occurred around every 4,500 years.

The intensity of the rain had increased by almost a fifth (19 percent). A rise in intensity of around 7 percent would be expected from global warming alone. That it is almost three times higher can't be explained solely by the fact that warmer air absorbs more water. In fact, the entire meteorological situation had changed. This probably means that storms like "Bernd" move more slowly, and there may simply be more of these storms. Which of the two this was (or whether it was a combination of both) wasn't part of our study, but our results did show that it would have rained less without human-caused climate change, and it may well be that fewer people would have died and more houses would have been spared. So the events in the Ahr Valley were no purely natural disaster, not least because the weather was influenced by climate change. But the unusual weather situation also played a significant role, and so the disaster would probably still have happened without climate change.

According to the International Disaster Database (EM-DAT, see chapter 3), more than two hundred people died in the German and Belgian floods of July 2021.[7] Houses, roads, and cars were destroyed, and the damage totaled tens of billions of euros. A small proportion of this was paid for by insurance companies, the rest by German taxpayers. The government put together an aid package of 30 billion euros at a speed usually only possible during elections.[8]

But how was the weather able to wreak such havoc?

DEVELOPMENT

The flood occurred in a region with a long tradition of human land use and the typical changes made to the surface through hillside development, drainage channels, and monocultures. All of these factors have an unfavorable impact on flood dynamics. A mapping campaign carried out after the disaster[9] identified systematic signs of erosion. Along small streams in particular, a sometimes above-average amount of debris, soil, and rubble slid down the slopes. The overall pattern seemed unusual. Wooded meadows with strong erosion were found next to practically untouched meadows and forests, and sections of intact slopes suddenly showed severe destruction. In most cases, satellite images identified the cause to be underground irrigation systems with pipelines. The satellite measuring devices (which process both visible and longer-wave light) spotted systems invisible to the naked eye.

Interviews with residents revealed that this "flagstone drainage" was common practice in the area to stop vineyards from getting too wet and dispose of the wastewater from the houses. These measures had devastating consequences during the flood of July 2021: as long as the drainage held, the mass of water flowed down the slopes and into the surrounding streams. Gradually, however, the overloaded drainage pipes filled with debris and became congested; the water backed up and the ground became so wet that landslides started to occur—a vicious circle in which flood waves started to build. Many of these drainage systems are several decades old and some aren't even documented, meaning that these potential sources of danger aren't included on flood maps or in the warning systems offered by the German Weather Service and the Federal Institute of Hydrology (Bundesanstalt für Gewässerkunde, BfG). Mapping the pipes, repairing or removing them, and renaturalizing the meadows will be a challenge, but this is urgently required to prevent similar disasters in the future.

The unprecedented damage in the Ahr Valley and the Lower Rhine Bay isn't just due to the sheer size of the area affected, but also to the high speed of the destructive water, which carried many trees along with it and produced tons of debris. This led to a flood level of several feet, flooding not just basements and first floors, but even the upper floors of many houses. Earlier severe floods in the Ahr Valley, such as those of 2003 and 2016, also caused significant property damage but weren't life-threatening. What made the 2021 flood so dramatic was the unfortunate interplay between extreme rainfall made more intense by climate change, the specific topography of the region, extensive impervious surfaces, and lots of falling debris as a result of clogged drainage systems.

Since the disaster resulted from a combination of unfortunate circumstances, nobody was immediately obviously in the wrong, and yet this was a case of injustice. Those affected were not given sufficient warning and the blocked drainage systems made the situation even worse, at least in part. An ongoing investigation[10] aims to clarify whether investments in robust early-warning systems were delayed, whether the residents were misled about the state of infrastructure provision, and why there were no flood prevention plans for smaller streams.

RISK AWARENESS

Even in a rich country like Germany, with high constitutional standards and functioning social systems, it was the people who couldn't help themselves who ended up dying—despite the floods being forecast. Days before the event, weather forecasts showed a high chance of flooding in the regions concerned. More than 150 public warnings were broadcast via the German Weather Service, European authorities, sirens, radio, and TV.[11] And yet residents and some authorities still believed themselves to be safe because they lacked any awareness of the risks, despite previous disasters (see chapters 2 and 3). An online survey conducted in the regions affected between August and October 2021 found that of 856 residents warned, only around a seventh (15 percent) expected massive levels of damage and life-threatening situations based on the warnings given. Most expected that all they would have to do was move valuables to the upper floors of their homes, while almost half (46 percent) didn't know how to help themselves and a third (29–35 percent) of respondents hadn't even noticed the warnings.[12]

There were no targeted warnings or recommended actions for children, the elderly, or people with disabilities, despite experience showing that these are the groups most vulnerable to flood events; they are less aware of the risks, struggle to remove themselves from dangerous situations, and are often harder to reach.[13] The town of Sinzig by the Ahr River suffered a tragedy when almost all the residents of a home for disabled people drowned after all forms of help arrived too late.

Particularly vulnerable groups fell victim to Germany's lack of awareness of the risks of weather disasters. The heat waves of 2003 and 2010 were two of Europe's deadliest extreme weather events of the twenty-first century.[14] Both were extremely well forecast, but hardly anyone knew where to start with the warnings. We could have and should have learned from these events, especially as we've known for years that advancing climate change could make weather phenomena more extreme. Nevertheless, even today Germany hasn't introduced any educational measures to build risk awareness among the population, not in schools nor in workplaces or care homes.

The disastrous flooding of 2021 also revealed gaps in official communication structures. In a federal system like Germany's, the national government is responsible for civil defense, but the states are responsible for disaster protection. This shared responsibility meant it wasn't clear which authorities were supposed to warn the people or how. And so Germany learned from bitter experience that warnings have to be practiced.

Bangladesh has viable communication structures that show how many lives can be saved by the appropriate communication of information, even if the events are new to the population. Instead of flood warnings just being broadcast on radio and television, critical information is sent straight to people's cell phones. Personalized warnings focused on consequences— like telling people not to enter certain slums in Dhaka because heavy rain presents a danger to life—can help to protect people in affected regions. If you don't understand a warning, you can't respond to it. So it's crucial that warnings are as clear as possible; people don't need to know that ten millimeters of rain are expected, they need to know about road closures and flood risks in specific districts or regions.

But all this will only work if people in institutions take responsibility and if investments are made in communication, education, and early-warning systems. In Germany, this is certainly compounded by the fact that after the Cold War ended, investment in civil protection was massively scaled back and authorities conducted hardly any drills. After reunification ended the permanent nuclear threat to the population, the country was lulled into a false sense of security. But now, the relevant people—the ones with the power, money, and other resources—need to take seriously their duty to provide a fair society.

WHO'S TO BLAME?

The question of guilt is hard to answer when it comes to weather disasters. Who's responsible for mapping the wastewater pipe network in the Ahr Valley? The authorities, or the people who laid the pipes decades before without considering potential risks? These are the aspects that occupy investigation committees and courts. This definitely makes sense, particularly in instances of individual negligence, but it doesn't help to resolve the underlying issues of structural fairness.

Injustice isn't always wrongdoing. The distinction is often blurred and depends on how judges interpret the law. For example, if building regulations are circumvented to build homes on the cheap, poor people will live at potential risk from heat waves. The same goes for building regulations that are insufficient to protect their inhabitants not just from rain, but also from storms. In both cases, the residents are victims of injustice. But if nobody is aware of how building regulations need to be phrased to provide optimal protection for people and the environment, or if buildings are much older and therefore hard to renovate, then it's not clear whether injustice has been committed. It's definitely unfairness. Blame and responsibility are different things. Elected politicians are responsible for developing building regulations, but do not bear any legal blame. Establishing justice is extremely difficult when there is no clear apportionment of blame, and yet it is absolutely necessary. And we mustn't forget that the media has a responsibility too. As Greta Thunberg says, "Our inability to stop the climate and ecological crisis is the result of an ongoing failure in the media... A crisis of information not getting through—because that information has not been told, packaged or delivered as it should be."[15] This is one aspect of blame for the climate crisis, along with journalists who actively spread climate lies (see chapter 7). But what about those who just ignore the topic? Are they guilty?

Or, to put it another way, do individual journalists have a responsibility? Yes! Editorial structures and how projects are assigned may mean they don't have the power to fully discharge their responsibilities, but they definitely have a responsibility. Just like the authorities in the Ahr Valley, even if they may not shoulder any legal or moral blame. In federalism, the unclear distribution of responsibilities between national and state governments can mean that responsibility is pushed back and forth between individual authorities and reappraisals fizzle out in the face of mutual recriminations. Why did the evacuations not begin until late evening when the water had already flooded entire houses, particularly in the Ahrweiler administrative district? Who is responsible for many residents having to fight for their lives on the roofs of their houses or even drowning? These are important questions, and the answers will help to mentally process the disaster, but ultimately they won't create a more just society. Justice can't be separated from responsibility, but responsibility can be taken without an admission of guilt or demonstrable blame.

If we can internalize this, then multiple levels of government can actually help us to take responsibility unilaterally.

One state could turn the messages from the weather service into personalized warnings for the local population (for groups, not individuals) and send them directly to cell phones without first having to go through laborious government administrative processes. Other federal states would be sure to follow this example.

STRUCTURAL RACISM

Field laborers in The Gambia don't have the same room for maneuver as white, wealthy Europeans when dealing with the personal impact of weather-related disasters. In this case, injustice takes on greater dimensions in The Gambia because in addition to structural sexism and often-inefficient administration, there remain postcolonial patterns of Western exploitation. Stories of the "poor victims" on the African continent are just one facet of African countries that we like to present in the West. Ultimately, however, the image of a dysfunctional continent is born from the same attitude that ships Europe's hazardous waste to Ghana. This is a racist attitude, an attitude of "us versus the others."

Back to my geography teacher, who—embracing this mindset—proclaimed Europeans' (weather) privileges to be "natural" in origin and nothing to do with centuries of power practices deliberately applied and maintained through repression. The bitter irony of this story is that this attitude has made many Europeans blind to the vulnerability on their own doorsteps, in their own villages, their own municipalities, their own country. What my lessons taught me is that it is "the others" who suffer from natural disasters.

As a result of this misconception, Germany trails behind other countries in the matter of climate justice for its own people. And it doesn't just trail behind France, which, after the dramatic heat wave of 2003, continuously invested in solutions and early-warning systems (Germany only began to design heat plans in 2023); it also trails behind many Global South countries, which offer examples of how to protect the population at a municipal level. The Indian city of Indore in the increasingly drought-plagued state of Madhya Pradesh has begun to renaturalize bodies of water and restore historic wells. Its many independent springs have helped reduce the burden on the existing water supply system and improve water retention in the ground. This measure has made the population far more resilient in times of drought (particularly in more informal settlements).

Climate justice means protecting the rights of those most at risk by sharing the burdens of climate change and its effects fairly and across all sections of society. For the most part, these burdens and effects are reserved for the Global South alone or for future generations, not for the West German wine regions affected in summer 2021. But we have to talk about climate justice in connection with events like these as well. Germany's homogeneous population (compared with other mid-European countries like France and the United Kingdom) certainly contributes to the lack of awareness of structural racism.

In a society like Germany's that remains patriarchal and racist, administration is oriented toward the needs of relatively prosperous white men, most of whom consider themselves relatively invulnerable. While topics like diversity and protection of minorities are now unavoidable at a national level—even in Germany—the situation is often very different at a local and regional level. In the previous chapter, we looked at how, in Australia, the narrative fueled by the fossil-fuel industry is questioned less in rural areas than it is in cities. It's the same in Germany, and Germany is extremely rural outside of its few large cities. In this context, "rural" doesn't mean agricultural;

it means a comparatively low population density with little immigration and emigration, and thus very little change. In these regions, local politics tend toward a racist and patriarchal mentality.

The only remedy is to *Unlearn Patriarchy*, as per the title of the German anthology by Naomi Ryland, Lisa Jaspers, and Silvie Horch. What does this mean? It means exposing and shedding the patriarchal patterns of thought and behavior that are embedded in our society, for the most part unconsciously and without reflection. The concept of decolonization is interesting in this context; to decolonize is to scour societal structures, laws, work processes, and other socialization mechanisms for colonialist and racist influences. There is still little awareness in Germany of the need for decolonialization, partly because Germany is not generally aware of its colonial past (in contrast to the Holocaust, for example) nor has it even begun to take responsibility for this past. But more people may become willing to do so if disasters like that of the Ahr Valley show that both the sense of invulnerability and the unquestioned sense of entitlement to privileges are incorrect models of thought rooted in colonialism.

The persistent notion that the participation of all "minorities" in public life would chiefly benefit these minorities—and not the majority of society—is another throwback to times gone by. I'm not saying that nobody would have died in the Ahr Valley if Germany had given more than lip service to diversity. But we can't simply dismiss the idea that a community aware of the diversity of its people will take measures to protect different groups differently according to their specific needs.

COLONIAL-FOSSIL ROOTS

What does all this have to do with the question of guilt and responsibility for the events involving the Ahr and Erft and for protection against further disasters? Individuals and authorities have failed, but what I find even more dramatic is the failure of our Western societies to recognize what climate change actually means. We don't have a technical issue; we have a global justice problem, and for that we can thank the prevalence of the persistent colonial-fossil narrative.

This doesn't just apply to the disastrous flooding of 2021 or the heat waves of the years before (and those that await us in the near future), but also to discussions on adhering to the 1.5°C (2.7°F) target. We won't achieve this target unless we break the spell of the colonial-fossil narrative. The colonial-fossil world order, and with it Germany's economic and social order, is based on the exploitation of workers and ecosystems to enrich a privileged few.

Whenever Greta Thunberg, Luisa Neubauer, and other activists make public statements on this matter, the impulsive backlash is by no means limited to trolls. In March 2019, Christian Lindner, Germany's federal minister of finance (of the Free Democratic Party) wrote on X (then Twitter) that climate protection was a matter for professionals, implying that the activists didn't know what they were talking about and wouldn't understand the contexts.[16] His tweet echoes the mood among broad swaths of the population, who instinctively reject a rapid transformation of the economic system.

The Climate Book (published by Greta Thunberg in 2023) received similar feedback. In this book, scientists, activists, journalists, and literary figures present the causes and consequences of climate change and explain the measures that need to be taken right away. A critic for a popular public radio station called the book infantile and claimed that it fell short by not including a section on technology.[17] This reaction is a good example of deliberately misunderstanding the issue; the book's core message is that climate change isn't a problem that can be solved with technology. People like to portray activists as naïve in this context, claiming that they don't know the social and economic realities. Thus interpreted, to be "infantile" is to depict the colonial-fossil narrative as unjust and in urgent need of change. This doesn't deny climate change, but every form of real solution is vilified as naïve and made to look ridiculous.

Another practice that supports the narrative of the fossil-fuel industry is the everyday, somewhat unconscious continuation of the habitual and learned. We simply carry on as we always have done. Day-to-day stress is used as justification for our failure to reflect on our own sexism and latent racism, let alone our own colonial-fossilism. We need to fill that position at work as quickly as possible; there's no time to think about ways of attracting underrepresented applicants. That new colleague's name is so complicated, we'll just call them something that sounds vaguely similar; of course it was an accident when the minutes of the meeting were sent to the wrong email address. The children are so fussy with their food; there's no time to think about where it comes from. Clara's ballet class starts at 4 PM and Paul has to be at football at 4:30; we can't start getting the bikes out now.

CHANGE

For most of us, life is stressful enough that just getting through the day is a challenge. And actually, taking the bikes out as such isn't going to change anything. But we do need to find the strength to tell a different story about how to live a good life in our society. Can we do this? And if so, how?

I have used the German flood disaster of 2021 to try to illustrate the actual problem and discuss how we can move from blaming individuals to social responsibility. How can we achieve widespread recognition of the harm that the colonial-fossil narrative is causing to all of us?

It's no coincidence that I'm addressing these aspects in a chapter about Germany. This is the country I grew up in, the place where I learned the tools I now use in my science. It's also the country I left. I didn't initially leave with the intention of never returning, but I would find it difficult to live in Germany today, as I wouldn't be able to conduct research as I do in the United Kingdom. So I may not be the right person to ruminate on how German society ought to change, and yet this transformation is a cause close to my heart. And I know that the generation of Greta Thunberg—both in the U.K. and in Germany—sees climate change as an issue of justice. That takes courage! But it seems that most voters and decision-makers in both countries (and beyond) can't or don't want to understand what it's really about.

Only when the public is made aware of the problems with the dominant narrative can it also be changed. And I'm not the only one looking for the right way to do so. The Canadian writer Naomi Klein, French economist Thomas Piketty, American writer Maggie Nelson, my friend (and YouTuber) Dr. Adam Levy, and groups like the Last Generation (Letzte Generation), Just Stop Oil, and Extinction Rebellion—they all (and many more) are talking about the best ways to get people to change their behavior, be it at the ballot box, in the supermarket, at the gas pump, but, above all, wherever they pass on narratives. Should we employ Naomi Klein's "shock doctrine," which favors focusing on the disasters fueled by climate change, or do we need a totally new concept for "capital in the twenty-first century," which Thomas Piketty is developing in relation to social inequality?

Activist ideas and practices are vital for new narratives, alongside these academically inclined concepts. But some of these ideas and practices are to be enjoyed with caution, as they unwittingly continue the fossil narrative; they stem from certain traditions in the environmental and peace movement that are markedly patriarchal and sometimes even serve a colonialist function. One glaring example is Extinction Rebellion, whose cofounder Roger Hallam described the Holocaust as "just another fuckery in history" (later stating that his remarks had been taken out of context).[18] In chapter 5, I cited Lazenya Weekes-Richemond and her analysis of the toxic and colonialist structures of many NGOs that have actually taken up justice as their cause. There is a similar danger with climate justice as soon as activist movements fail to reflect on their own traditions. Well-meaning concern can't be addressed with the wrong methods, and if this results in toxic climate justice, it will benefit nobody and harm everyone who wants to effect change. One example of toxic climate justice is when activists publicly discredit scientists for flying to conferences. Or when people with several children are judged for bringing people into the world who, statistically speaking, will have high carbon footprints in the future.

To actually achieve climate justice, we need something very special: new narratives that aren't aligned with the traditional understanding of justice and are incomparably more attractive besides, narratives that everyone can open up to without having to think about it.

Is that possible? Of course it is! Extremely powerful narratives have been changed throughout human history. Slavery has been abolished, people of color have full civil rights, gay couples are allowed to marry and raise children, and women participate fully in society. Today these things seem like a matter of course, but less than a century ago entirely different narratives were at play, and the sense of justice was entirely different. We still haven't eliminated sexism or racism (as discussed), but the overall picture has changed. This was achieved through protests, education, art and culture, and by changing power structures. The American writer Maggie Nelson makes it clear in her books that nobody comes into the world knowing about unequal power structures and their impact. Injustice is learned, and that means it can be unlearned. And so we absolutely need protests, education, and art that address climate justice. And we need more of them all! "Climate fiction" may now be an established term, but I've yet to read a truly good novel on the subject. And I still haven't found the visual arts I'm looking for. At least that's my personal experience. Maybe I need to search more thoroughly, but this too shows that the topic simply isn't prevalent enough.

Ultimately, power structures are the main problem. If humans found it easier to relinquish power, the prevailing narratives could have been changed much faster. There are so many areas in which changes are urgently needed today, all of them important: education, money, family, science, identity, sex, work, racism, art, and culture.

We won't achieve climate justice with an approach based solely on damage, but this must be one of the pillars of its development. We should keep an eye on emissions, temperature developments, and their repercussions with scientific precision. Disasters like the 2021 floods in the Ahr Valley must be reviewed in detail with meaningful consequences, and this is already happening.

The reform of Germany's disaster protection systems was accelerated during the Covid-19 pandemic with added pressure from the disastrous flooding. Relevant agencies received more funding for staff and equipment. A new situation center was also set up that brings together stakeholders from the federal government and state departments of the interior and enables better coordination of crisis teams at federal and state level. Two months after the disaster, a funding program was launched for electronic warning sirens that can also be activated by federal authorities directly. The government's official danger warning app (NINA) was reworked, and the cell broadcast warning system was introduced, which sends messages directly to smartphones in affected regions throughout Germany. There are also plans to overhaul civil protection. Compulsory training and regular drills for the population are being discussed along with ideas on how to recruit civil protection volunteers.[19]

These are all desirable measures, but we must not lose sight of the fact that climate justice is being impeded by the reproduction of power structures that Max Liboiron (a Canadian geographer and specialist in civic involvement and environmental pollution) calls "neoliberal" and that I call "colonial-fossil."[20] Over two centuries, the world has been squeezed into a corset of colonial and fossil growth. To clear the way for climate justice, this corset must be ripped open and the mental barriers of the popular logic of inherent necessity (which portrays growth per se as salvation, regardless of who profits from it) must be cast aside. For this to happen, every single one of us should take responsibility for change.

9

A COUNTRY DROWNING IN CLIMATE DAMAGE

Pakistan

"I see hundreds and thousands of people, helpless people. I see a complete blackout in this area. There is no electricity here, and there is no internet connectivity. People are trying to call for help... And people... have refuged on some island or some paved road or any place where they could save themselves from water."[1]

"This is where the little one sleeps. All we have is the floor. We have nothing. I have these dishes. Nothing else. No food to cook. Everything's empty."[2]

"Let me be clear, this is about climate justice."

"We are not blaming anybody, we're not casting allegations, what we are... saying is this is not of our making but we have become a victim. Should I be asked to cast my appeal into a begging bowl? That is double jeopardy. That's unjust, unfair."[3]

A COUNTRY UNDER WATER

Mid-June 2022 marked the beginning of potentially record-breaking monsoon rains throughout much of Pakistan. The worst-affected provinces (Balochistan and Sindh) had seven or eight times more rain than usual; across the country, rainfall was around three and a half times higher than normal. The intense rain triggered large-scale flooding, landslides, and flash floods. The Indus River, which runs through the whole of Pakistan, flooded thousands of square feet, creating a lake over sixty-two miles (one hundred kilometers) long in the south.

These exceptionally strong monsoon rains fell on a country with a high population density and significant poverty. More than 33 million people (Pakistan has a population of 240 million)[4] were affected by the floods. Over 1.7 million houses were destroyed, and almost 1,500 people died.[5]

On August 25, the government declared a national state of emergency. The acute damage was estimated at around 30 billion U.S. dollars by the World Bank, not including the cost of rebuilding. It will take years for the region to recover. More than 4,350 miles (7,000 kilometers) of roads, 269 bridges, and 1,460 healthcare facilities were destroyed. Over 18,000 schools were damaged, almost 1 million farm animals were killed, and around 6,950 square miles (18,000 square kilometers) of farmland was ruined. Almost half of the cotton harvest—one of Pakistan's most important exports—was wiped out by the floods. Other crops worth around 2.3 billion U.S. dollars were also destroyed, further exacerbating the food shortages caused by the war in Ukraine and the extreme heat waves of the pre-monsoon season in the same part of Pakistan.[6]

There were outbreaks of diseases like diarrhea, cholera, skin and eye infections, and malaria, particularly in Sindh and Balochistan. The poor hygiene conditions in the emergency accommodation provided for refugees—estimated at 664,000 people in September 2022[7]—made the medical situation even worse.

VULNERABILITY

According to the latest ranking in 2021 by the NGO German-watch, Pakistan is one of the countries most at risk from climatic extremes. This is partly due to its geographic location between two major rain-bearing weather systems: the monsoon rains from the east and southeast in the summer and low-pressure systems (westerly disturbances) from the Mediterranean in the winter.[8] Studies show that climate change has a significant influence on both phenomena, but more on variability than on intensity.[9] This means that even if the average amount of rain remains about the same, individual years will be significantly wetter or drier. These changes in weather due to climate change make Pakistan even more susceptible to disasters like that of 2022.[10]

Pakistan had already experienced catastrophic flooding back in 2010. The Report of the Flood Inquiry Commission appointed by the Supreme Court of Pakistan concluded that this wasn't merely the result of extraordinary weather events; much of the damage was caused by the deliberate breaching of dikes and dams to protect large cities and industrial facilities.[11] Economically speaking, Pakistan is a poor country with few opportunities to adapt to weather extremes. It is currently ranked 150 (out of 181 countries) on the Notre Dame Global Adaptation Initiative (ND-GAIN) index, making it one of the worst-prepared countries in the fight against climate-change impacts. The index measures a country's adaptability based on economic, structural, and social criteria (the economic criteria include openness to innovations to mitigate climate and weather-related disasters). Institutional factors, social inequality, and educational support for adaptation measures are also taken into account. At the time of writing, Norway holds the top spot, and Chad is right at the bottom.[12]

The investigations that followed the great flood of 2010 also showed that strengthening existing dams and channel systems alone wouldn't prevent future flooding and a similar disaster could happen again. That's exactly what happened in 2022, and unless fundamental changes are made, it won't be the last time. There are two crucial reasons why cities and villages along the Indus River Delta are so vulnerable. First, there's the engineering-based river management method used by state planners, which makes water available to cities and industrial facilities as a "resource" without considering the needs of the region's smallholder farms. Second, the country places its trust in an irrigation system set up by British India for very different political reasons. The British occupiers wanted to demonstrate their might, so their structures had to be as imposing as possible. Accordingly, they established an irrigation system that then created other problems and didn't improve the water supply for everyone in the region.

Pakistan later also mainly favored hydrological megaprojects to manage flooding. But the huge dams that have been built have led to a range of problems; entire regions have dried up as a result of large-scale intervention in natural river courses. As this infrastructure began to age, severe sedimentation caused further problems, reducing channel capacity and fostering flood conditions.

Since the early twentieth century, flooding has occurred along the Indus River due to deliberate breaching of dams, which diverts the flood water. Pakistan's current susceptibility to flooding is therefore also the result of colonial and post-colonial politics. The practice of diverting excess water to the countryside to protect the cities has been used since British colonial rule and continues to this day. This is why the 2022 floods mainly affected rural areas, where many people (64 percent) earn their living from agriculture. Agriculture is an important part of the Pakistani economy and makes up 26 percent of its GDP.[13] The Indus River supplies water to almost 90 percent of Pakistan's food production, which employs half of the country's workforce and produces wheat, rice, and sugarcane.[14] These are the people who suffer the most when flooding occurs. In response, many farmers invested in cattle production, although this has not proven to be a sustainable way of protecting against dramatic losses. Large farm animals are particularly at risk from flooding. It is estimated that several hundred thousand animals drowned in Sindh province alone in the 2022 floods.[15]

LOSS AND DAMAGE

In the same year that Pakistan flooded, the COP27 climate change conference was held in Sharm el-Sheikh, one of Egypt's most popular coastal resorts, at the southern tip of the Sinai Peninsula. The conference was sponsored by Coca-Cola, which cynics might describe as real progress, given that the Polish coal industry had stalls at COP24 (2018, in Katowice).

The Pakistan floods were a tragic example of the importance of loss and damage, the focus of the 2022 conference. "Loss and damage" is one of the three pillars of international climate policy, alongside mitigation (preventing future greenhouse gas emissions) and adaptation.

These three pillars are the areas set out in the 2015 Paris Agreement and in which climate change has to be addressed politically. But a house built on these pillars wouldn't stay standing for long; the mitigation pillar is huge right now, but the adaptation pillar is thin and feeble, and the loss-and-damage pillar is decorative at best.

This imbalance is the result not of the Paris Agreement itself, but the way in which it is implemented.[16] For mitigation, the IPCC Working Group for greenhouse gas inventories had a comparatively large literature base with which to develop standards for measuring greenhouse gas emissions. This enables the progress achieved in reducing greenhouse gas emissions to be measured and the nationally determined contributions (NDCs) to actually be verified.

The situation is quite different for adaptation. The Paris Agreement aims to strengthen global capacity to adapt to the negative effects of climate change. Article 7 defines a global goal to improve the adaptability and resilience of cities and communities, to reduce vulnerability, and to contribute to sustainable development.[17] But unlike with mitigation, there are no measurable criteria here. Individual countries have defined national adaptation guidelines, but these have tended to be vague. For example, the German federal government committed itself to presenting a preventive climate adaptation strategy with measurable targets, updating this regularly, and ensuring its ongoing implementation, but then tasked the states with figuring out exactly what that was supposed to mean and submitting the strategies.[18] In addition, there are no comprehensive studies on how people actually adapt to climate change that could be used as the basis for corresponding metrics (qualitative or quantitative).

Adaptation measures must be conducted at a local level and include such varied aspects as strengthening national health-care provision, early-warning systems, new agricultural methods, dams, and awareness campaigns. Adaptation may be difficult to measure, but it isn't impossible. Here too, measurable factors could be agreed, such as warning exercises or the number of people with health insurance as benchmarks. But so far, there seems little will to do this. The global talks at COP26 in Glasgow in November 2021 made clear that there is a great deal of uncertainty around the global goal of adaptation and how relevant progress could be measured. And so no specific steps were agreed. But before Glasgow 2021, an agreement had been reached that from 2020, the Global North would make 100 billion U.S. dollars available every year to support adaptation measures.[19] Payments of this amount have not yet been received, but at least the growing deficit can be measured.

The situation is even trickier for loss and damage; there isn't even a common vision of what is to be achieved. Article 8 of the Paris Agreement simply states that "parties recognize the importance of averting, minimizing and addressing loss and damage associated with the adverse effects of climate change, including extreme weather events and slow onset events, and the role of sustainable development in reducing the risk of loss and damage."[20] Back in 2013, the so-called "Warsaw International Mechanism" was deployed, a still-vaguely defined instrument that is supposed to help overcome loss and damage. It is unclear exactly what this involves and what it is supposed to do. Since the Paris Agreement was enacted in 2015, there have been many debates on the ominous Warsaw International Mechanism; sadly, no real progress has been made as yet— which is why establishing a specific fund (as happened in 2022 and was confirmed in 2023) is a huge step forward. But it's still far from providing any vulnerable country with serious help.

Implementing the fund is a struggle—among other things, because loss and damage is not defined in the context of climate change. The current wording allows for anything from specific economic losses due to rising sea levels or extreme weather events through to exclusively precautionary measures indistinguishable from adaptation strategies. How are you supposed to tackle a problem when it isn't even clear what the problem is? This dilemma came about because the topic was added to the political agenda at the insistence of island nations like Mauritius, the Bahamas, and Fiji. These nations are confronted most strongly with the reality of climate change but have played practically no role in the causes. But thanks to the Global North countries, the definition of damage was left deliberately vague. So progress has been made, at least on paper. Success can be sold to the voters back home, but in reality this doesn't change anything. Even prevention—which has specific, measurable targets—is no different. The German government has already been successfully sued because its laws do not fit with the goals of the Paris Agreement. And yet even in 2023, it didn't have any climate laws that were being complied with and that were compatible with the global targets.

THE JUSTICE PROBLEM
IN A NUTSHELL

Loss and damage is clearly linked to justice; the basic idea behind the concept, as brought forward by low-lying island nations, is for the Global North countries to take responsibility as the main contributors to climate change. This was ignored for a long time, particularly as the lack of definition meant it wasn't clear exactly what was supposed to be negotiated. Although pressure grew year by year, not least due to the ever more obvious damage, the parties involved found themselves in a conflict of interests: the Global North countries don't want to pay for the impact their colonial-fossil policies have had in previous years, and the Global South countries have no interest in making compromises that, in the case of smaller island nations, might threaten their very existence. This doesn't mean that these islands will disappear completely, although that is certainly possible over longer time scales of a few hundred years. What it does mean is that the combination of rising sea levels and fierce cyclones will lead to extensive flooding, causing the soil to become salty and removing the possibility of sustainable reconstruction. This won't just destroy areas in which people live and work; it will wipe out ancient cultures and traditions.

One initial success for the highly vulnerable countries was the inclusion of loss and damage in international climate agreements as a topic for negotiation.

Scientific surveys were conducted among the negotiators at climate conferences to find out how loss and damage could be defined. These surveys showed the huge divergence in opinions on what loss and damage means and what should be done about it at an international level.[21] Most of the people interviewed agreed that the regulations to be developed should address slow changes (such as rising sea levels) and extreme weather events. There were also similarities in opinions on the time scales of the relevant measures. Most interest groups agreed that future damage should be prevented (*ex ante*) and loss and damage already incurred should be regulated (*ex post*). However, opinions varied greatly on how these should be weighted—and on other aspects too.

One group of negotiators favor an interpretation of loss and damage that largely corresponds to the adaptation strategies. They emphasize that the mandate of the UNFCCC (UN Framework Convention on Climate Change) is to prevent climate change, implying that this is just about the future, and the adaptation instruments already available are sufficient to deal with loss and damage already incurred. It's hardly surprising that this rationale comes mainly from the American, Australian, and European negotiators, who ultimately say that the difference between damage and adaptation measures lacks any scientific foundation and is purely political.

This contrasts with the interpretation of loss and damage as regards the limits of adaptation, both in the past and in the future. This interpretation is aimed at damage to which we cannot adapt—for example, if an island becomes uninhabitable as a result of rising sea levels—and damage to which humanity cannot adapt because of a lack of political will and/or the requisite knowledge and capital. However, most negotiators who take this stance don't want to make a distinction as to whether this loss and damage has anything to do with climate change. This makes sense insofar as it helps those nations (and, above all, their people) who are highly vulnerable and have low emissions. While this position ignores everything that attribution research can do, it saves time in providing assistance. Another argument in favor of this approach is that we can't perform an attribution study for every extreme weather event; suitable models aren't always available, nor is there always enough observation data. Nevertheless, this leads to another problem, which risks undermining the Paris Agreement in the long term. If we leave aside climate change as a cause of loss and damage, then why should this category even be included in the Paris Agreement? The regulation of loss and damage would then be nothing more than development aid, and it could easily be argued that it wouldn't need to be included in climate talks at all. For this reason alone, it seems wrong to me to introduce a mechanism within the UNFCCC that doesn't distinguish between damage caused by climate change and damage caused by something else.

But perhaps I'm falling into the same trap as all those who want to see climate change treated as a purely scientific phenomenon, and overlooking something crucial: if we were to focus solely on the loss and damage resulting from climate change, wouldn't that ultimately entrench the injustice that led to climate change in the first place?

The analysis of Pakistan's dramatic flooding sheds light on this issue.

ATTRIBUTION

The devastating Pakistan floods of 2022 made headlines around the world. On the BBC, the event was described as a "monsoon on steroids" and massive "glacial lake outbursts," while *The Guardian* talked of "record-breaking rainfall."[22] Sherry Rehman, Pakistan's minister of climate change, also attributed the floods to climate change and called on the Global North to pay reparations—difficult not least because there is currently no clear way to demonstrate how high these reparations would have to be and the share that climate change has played in this event.

We showed why this is not a straightforward problem in an attribution study conducted in October 2022 together with the Pakistani weather service: the floods proved to be a direct consequence of the extreme monsoon rains of summer 2022. In August 2022, the provinces of Sindh and Balochistan hit extreme peak rainfall values, which exacerbated the situation further. We therefore examined the sixty-day and five-day maximum rainfall for the entire Indus basin and both provinces during the monsoon season.[23]

Pakistan is located at the far western edge of the South Asian monsoon region, with a predominantly dry desert climate in the southern provinces (this includes Sindh and Balochistan) and a wet climate in the north. There is often heavy rainfall from June to September, particularly in the north, and over half a million people are affected by floods each year. However, this is normally caused by local low-pressure systems, rather than those that develop during the monsoons. Pakistan gets far less rain on average than the Indian states of Uttar Pradesh and Assam, which are at the same latitudes. Rainfall also differs greatly within Pakistan: some years it hardly rains at all, and in others there are dramatic floods. This is why multiyear weather observation data is required to gauge what is "normal" weather and what is actually extreme.

In the midlatitudes, like in London, rainfall may differ from day to day and week to week, but yearly comparisons remain similar; in addition, the U.K. has been keeping weather records for longer than Pakistan. Here—in one of the countries that, historically speaking, is primarily responsible for climate change—we can use both the local climate and long series of observations to calculate pretty accurately which events take place once a century, for example, and how much climate change influences the intensity of weather events. This information can then be used to deduce very precisely what additional damage has occurred and the level of losses.

In the case of the Pakistan floods, the available weather data shows that an event like the extreme rainfall of 2022 can occur every eight years at most; in purely mathematical terms, it's also possible that it will never happen. The reality falls somewhere between these two possibilities—but where exactly it falls can't be determined with the available data due to the high variability of the weather. It's also difficult to calculate the proportion attributable to climate change. Statistically, we can't rule out the possibility that it plays no role at all, but it's equally possible that climate change has made this extreme event up to one hundred times more likely.

But statistics aren't our only source of information. Physics tells us that warmer air can absorb more water vapor and therefore leads to more extreme rain. Studies show, at least over short time scales (one to five days), how the intensity of monsoon rainfall increases with global warming across South Asia.[24] We can therefore say with great certainty that climate change did play a role in the flooding in Pakistan, but we can't quantify its influence based on the information available. That was the (not especially satisfying) conclusion of our attribution study.

Although some of the loss and damage was the result of human-caused climate change, it is not currently possible to quantify the specific amount. This is largely due to the climatic conditions and the comparatively short observational weather datasets—which, incidentally, are comparatively long in Pakistan. In parts of West Africa, for example, where the climate is often even more variable, weather records are very sporadic, making it even harder to characterize extreme events and attribute damage to climate change.

In the countries that have barely contributed to climate change, and where vulnerability is particularly high, loss and damage can barely be quantified. So if a mechanism were to be introduced at an international level for loss and damage caused by climate change, there would be a danger that it would fail in its objective and put low-income countries at further disadvantage.

CALLING FOR JUSTICE

There's another problem with the Paris Agreement text when it comes to fulfilling the call for reparations for the damage caused by the Global North in Pakistan and elsewhere. The "loss and damage" category doesn't imply compensation payments; in fact, it actually excludes them. That "loss and damage" doesn't explicitly mean compensation[25] is down to the pressure exerted by the Global North. This definitely deserves criticism and makes clear the urgent need for a precise definition of loss and damage that can be used as a foundation for actually beginning to address the issue in the real world. Is this just another name for help with adaptation, or is it more a reflection of the call for justice and climate reparations? Right now, my answer is probably the same as even the most involved negotiators: I have no idea. And despite smaller-scale negotiations throughout 2023, there remains a lack of clarity. It's agreed that reconstruction aid is important, but so far there's nothing more specific than that. And it probably won't get more specific until money is actually being paid (and the 700 million U.S. dollars promised at COP28 in Dubai, just 2 percent of the costs of the Pakistan floods alone, is not actual money in that respect). Loss and damage was the major topic of the 2022 climate talks, and at the UN General Assembly of the same year, UN Secretary-General António Guterres emphasized that this topic is a "fundamental question of climate justice, international solidarity and trust," adding that "polluters must pay" because "vulnerable countries need meaningful action."[26]

The talks culminated in a decision to set up a fund to help highly vulnerable countries deal with loss and damage. At the UN, money is the main thing. Money that the Global North doesn't want to pay, money demanded by the least-developed countries taking part in the climate talks—money they desperately need. Pakistan cannot overcome the consequences of the 2022 floods without financial support, just as Nigeria cannot pay out of its own pockets for the damage caused by the dramatic floods of the 2022 rainy season. But it's not just about the monetary side of things. No money in the world can revive the people who died in South Asia's 2022 heat wave or take the Bramble Cay melomys off the list of extinct species.[27] These losses, described in the literature as "noneconomic loss and damage" (NELD), are the truly dramatic consequences of climate change. Alongside countless deaths and extinct plants and animals, losses include the annihilation of entire cultures such as that of the Torres Strait Islanders, whose traditional burial sites, meeting places, art, and homes are gradually falling victim to rising sea levels.[28]

A THOUSAND FORMS OF LOSS

In one of the few studies that systematically looks at noneconomic loss and damage, the authors write that every piece of research on loss and damage caused by climate change must take people's lived experiences as its starting point. The focus should be on which losses occur where and how, how much they hurt the people involved, and whether they can be classified as acceptable and potentially reversible.[29]

The study undertakes a systematic, comparative analysis of climate-related loss and damage (regardless of whether this was caused by anthropogenic climate change). Using over a hundred case studies from around the world on the various causes of damage (such as heat waves, droughts, and rising sea levels), the authors show how these losses affect people's daily lives. The study also examines valuables that have been threatened or completely extinguished by the weather or climate change. The authors analyze examples of losses already seen around the world. These include tangible things like physical and mental health or a person's farmland, as well as intangible things like identity and cultural techniques. Strikingly titled "One thousand ways to experience loss," the study is intended as a wake-up call highlighting the moral and ethical contours of loss and damage and the notions of climate justice contained therein. This also highlights the urgency of supporting the weakest as they experience loss and of avoiding future damage with suitable adaptation measures. The authors conclude that reparations should be ensured for those affected to actively promote social solidarity and cultural practices.

The study shows how wrong it is to depict intangible, noneconomic loss and damage as random, trivial events, to shrug them off as a load of fuss that distracts from quantifiable factors that can be tackled with political measures (from top to bottom).

The examples used are very real: herds of Lapland reindeer dying from new parasites, male loneliness in Siberia and Nigeria when women leave farms to work in cities because the farmland has been destroyed, alcohol abuse and violence as a reaction to melting sea ice that makes traveling and hunting impossible, and the trauma of survivors forced to watch people and animals die by flood or fire. These aren't isolated cases; these are inherent losses that don't show up in insurance policies and aren't included in cost-benefit calculations. The scale of these unquantifiable losses may be infinite, and the fact that they are difficult to measure doesn't mean loss and damage can't be weighted.

The first thing to do is to genuinely recognize that social and cultural structures are being destroyed by climate change. This would also include admitting that climate change is one of the consequences of the colonial-fossil lifestyle, which is based on the exploitation of the many for the benefit of a few. This is also the basic motivation for incorporating loss and damage into UN talks. Ultimately this is about money too, the only currency that signals true acknowledgment and that can be used to fund projects to reduce vulnerability and to assist reconstruction.

COLONIAL AND PATRIARCHAL LEGACY

The inclusion of effective measures on loss and damage hasn't exactly been successful so far, and not just in UN climate talks. The topic remains an almost unknown quantity in science too. The IPCC publishes regular reports on the latest state of knowledge on human-caused climate change. To produce these reports, scientists examine, interpret, and summarize all scientific publications on the topic in a meta-peer review.

Accordingly, these reports can only make statements on contexts that have been researched academically. But the phrase "loss and damage" hasn't been used in any scientific studies. It doesn't appear at all in the report on the physical science basis.[30] This is partly because most scientists see themselves as "neutral," removed from political contexts; to me, this is an illusion. And so many researchers tend to exclude content with political connotations (like loss and damage) from the outset.

The second problem is the thoroughly colonialist and patriarchal structure of the scientific world and the IPCC itself. At first glance, the ratio of Global South authors (38 percent) to Global North authors (62 percent)[31] looks promising, but only if your goal is parity. World population distribution should be our yardstick—in which case the ratio would have to be around seven authors from the Global South to one from the Global North. And we're a long way from that.

Representation and, even more pertinently, equal participation would allow politically relevant science to be produced that is oriented toward real-life circumstances. But scientists from the Global South are often unable to participate because they need access to pricey scientific journals (access to all the academic journals published by Elsevier can cost over a million U.S. dollars, for example[32]). Plus, they need high-speed internet around the clock to attend meetings and conduct all the research they need. Many academic institutions in the Global South simply don't have the necessary resources.[33] Scientific literature is dominated by the Global North—not just in the number of papers submitted, but also in the underlying models and scenarios used in climate science. Not a single climate model in the world was developed in Africa, which ultimately means there isn't a single climate model *for* Africa.

The scientific bias from Europe and the U.S. becomes even more apparent when we look at socioeconomic future scenarios that make assumptions about economic, technological, and social developments and that are used as the basis for all

emissions and emission reduction calculations. The institutions that create these scenarios are in Germany, Austria,
Switzerland, the U.S., the Netherlands, Belgium, and Japan.
As a consequence, the emission reduction scenarios ultimately
make extremely problematic assumptions about the global
economy.[34] These scenarios assume high energy consumption and continuous economic growth by rich countries until
2100. Uninterrupted growth in the North is reconciled with the
targets of the Paris Agreement by assuming restricted energy
consumption in the Global South. These scenarios also speculate about technological changes to an extent not supported by
empirical literature.[35] The idea—also popular among German,
British, and other Global North politicians—that we can simply
invent technology to remove CO_2 from the atmosphere appears
so alluring to Western scientists that they too eschew scientific
realism altogether and employ technological wishful thinking
in their future scenarios. And this is despite numerous studies
clearly showing that this isn't possible within the time frame in
which emissions need to drop.

Even if these scenarios aren't directly relevant to the analysis
of loss and damage, they are symptomatic of the systematic disadvantaging of all those who have not historically profited from
the burning of fossil fuels. This shows just how much science
is stuck in the colonial-fossil narrative (particularly the natural
sciences, which pay far too little attention to ethical values).

But while there may be many challenges in defining and
measuring loss and damage, this doesn't mean the topic is too
complex to be addressed by science and politics. This topic
must form part of global climate protection policies because
ever-increasing loss and damage—such as that seen in vast
swaths of Pakistan in 2022—is threatening life as we have
known it for hundreds of years.

A BREAKTHROUGH?

At the 2022 climate conference, loss and damage was on the official agenda for the first time. At the end of the conference, a resolution was passed to set up a loss and damage fund that will make financial aid available to the most vulnerable countries. So far this remains nothing but rhetoric. The resolution simply stipulates that a fund is to be established; it remains unclear which countries are to pay into it, which countries will be able to access the funding, and how much money the fund will contain. Once again, all these questions have been deferred. The only thing set in stone is the appointment of a transitional committee that will clarify open questions on the establishment of a fund to regulate and prevent loss and damage.[36] This committee was then supposed to submit proposals for COP28, held in the United Arab Emirates in late 2023.[37] This conference had ended by the time this book was written, but aside from some minor agreements, many of the questions raised in November 2022 remained open. Once again, it was simply declared that action must be taken urgently and this requires money.[38]

Nevertheless, and justifiably so, the resolution to establish a loss and damage fund was lauded as an important breakthrough, because this does at least mean official recognition of the topic of loss and damage, a topic for which the Global North bears responsibility. Whether the Global North will fully accept this responsibility in the form of financial liability is more than doubtful, however, and when we look at the other resolutions of the 2022 climate conference (and practically every other conference as well), it becomes clear that international justice is not the primary goal of the U.S. and European negotiators. Other nations with vast oil wealth, such as the hosts of the 2023 conference, by no means hide their interests; Sultan Al Jaber,

CEO of the Abu Dhabi National Oil Company (ADNOC), was appointed president of COP28.

In the run-up to the 2022 climate conference, India attracted public criticism for watering down the strategy for the global fossil-fuel phaseout by speaking instead of a "phase down" (the gradual reduction of coal burning, leaving open whether it will be reduced to zero). But India changed course when the conference came around, perhaps inspired by its dramatic heat wave of 2022. On the first day of the talks, India called for the phaseout of all fossil fuels and received the support of many other countries, including the EU, the U.K., and even Australia and the U.S. The Global North managed to sneak a few restrictions back in by suggesting the word "unabated" for fossil fuels, which again allows people to bet on dreams of future technologies, rather than actually phasing out fossil fuels quickly.[39] At the end of the conference, no resolution had been passed on the phaseout of fossil energy sources. But while India had initially (and rightly) earned plenty of media criticism, it didn't receive any praise for its change of course, just as the U.S. wasn't criticized for altering the wording.

Media reports focused on the lack of progress in measures to adhere to the 1.5°C (2.7°F) target, while the failure to pass a resolution on the fossil-fuel phaseout garnered barely any attention. This can be seen as evidence of the persistence of the colonial-fossil narrative. The tale of fundamentally good fossil fuels may have started to fracture in recent years, but we are clearly still thinking within this framework. So the more than six hundred fossil-fuel industry lobbyists who took part in the climate conference have done good work at least.

There's certainly plenty of reasons to criticize Narendra Modi's Indian government, particularly when it comes to human rights and religious freedom. And yet India appears to have recognized the high price of lingering in the colonial-fossil framing and wants to drive change, at least at an international level. The same goes for Pakistan which, as the former chair of the G77 nations, played a significant role in getting "loss and damage" to be recognized at the climate talks with the intention of establishing financial rules. The events in Pakistan in 2021 and 2022 once again showed the world quite plainly the scope that loss and damage can take. The promise to set up a loss and damage fund was an important victory in the fight by all the people and nations that have not been historically responsible but have paid the price, like a large percentage of the Pakistani population. This fund must now be equipped with financial resources and made fit for purpose, despite all opposition and delaying tactics.

Some also doubt whether drawing so much attention to a politically highly problematic subject like loss and damage is a step forward for climate justice. This diverts attention away from a topic like adaptation, which is already given too little importance and for which the promised money still hasn't been paid.

The argument goes that if money is now promised to loss and damage instead, this further reduces the likelihood that adaptation measures will be financed. Looking at the last twenty-eight climate talks, the possibility can't be discounted. But I think this disregards the fundamental question of justice. When the aim is justice—changing the world, no less—it's important for loss and damage to be right at the top of the agenda, on an equal footing with adaptation and the transition to a carbon-neutral world. The goal is to give low-income countries sufficient financial resources, not just fob them off with verbal pledges. This is urgently required, as emphasized by the following comparison from Harjeet Singh, an expert on loss and damage with the Climate Action Network, who said that to "somebody who is drowning, we [the UN] say: 'We can't help you now, but if you survive we may help you prepare for a future disaster.'"[40]

Participants in global climate talks rely on consensus and collaboration, which is why it consistently proves difficult to arrive at common goals. We've gotten used to this, but if you think about it a little more closely, isn't it surprising? After all, shouldn't every country be able to agree that universal human rights are their greatest common goal? I'm not the only one whose day-to-day work highlights the extent to which climate change infringes on the human rights to life, health, culture, personal fulfillment, and water; UN climate conference delegates know about this too, as they are the ones who go through the IPCC reports with a fine-tooth comb.

And while there may have been progress in the twentieth century, colonialism remains present in the UN. This was spelled out very clearly once again by the battle over the distribution of Covid-19 vaccines. For the topic of loss and damage, this means that strategies for the inevitable damage of the coming years should be pursued outside of international climate talks as well. As long as calls for global climate justice remain nothing but words, with no actions to back them up, the nations affected will bear the brunt, largely at an individual level in households and local communities. But many people are now starting to resist this, with protests and in the courts, such as the lawsuit against RWE in Germany (see chapter 6).

In a country like Bangladesh, which often suffers dramatic flooding, people are very aware of the consequences of climate change and the historic responsibility of the Global North. In view of the sluggish progress made at an international political level in dealing with the loss and damage caused by climate change, Bangladesh is leading the way, becoming the first country in the world to set up its own program.[41] Meanwhile, the African Risk Capacity Group funds additional risks resulting from climate change using international aid (mainly from the U.K., Germany, Sweden, Switzerland, Canada, and France) rather than the premiums of insurance customers.[42] Incidentally, the German elected representative at the climate summit would have liked to push through this idea instead of a loss and damage fund. While this might have seemed a good solution at first glance, the topic of compensation and reparations would have been taken off the table for a long time, and the damage would have remained because climate insurance focuses solely on future damage, and not on damage already incurred. And even if we can certainly accuse a country like Pakistan of not having learned enough from the dramatic flooding of 2010, that doesn't change the fact that it is colonialism and human-caused climate change that caused the damage in the first place. To leave Pakistan to pay for this alone would not only be morally wrong and unjust; it would entrench an ever more unjust world in the long term.

10

WHAT NOW?

THIS BOOK IS about the weather and climate, but it's also about poverty, sexism, racism, arrogance, ignorance, and power. The examples I've offered are often harrowing, and that's despite only touching on war and mass migration, two of the most dramatic consequences of climate change. There are several reasons why I haven't explored these in more detail. Unraveling causal chains is often very difficult, as it's rarely clear whether or to what extent climatic changes play a major role in the outbreak of a conflict or a person's decision to leave their country. When climate change destroys harvests through extreme heat, for example, the people who rely on the crops for their income are left with no economic resources whatsoever. And you can't emigrate without money. As we saw in the chapter on Madagascar, in cases like this climate change actually reduces migration locally. On the other hand, people may emigrate simply because negative climatic changes are expected. The fact that the climate is changing, even if the effects can't yet be felt directly, is forcing many people and businesses to make life-changing choices—for example, driving wineries to Great

Britain and dairy farmers to other countries and professions. If things go well, this will result in a better, more successful life, but it can often be tricky to survive in a big city. As a rule, a choice as momentous as emigrating is made for many reasons; sometimes, people aren't even aware of all the factors that made them leave.

Researching the causes of war and social conflict is even more difficult. In May 2023, more than six hundred people died in Rwanda and the Democratic Republic of the Congo after heavy rainfall. At least that's how it appeared from reports by local media and humanitarian organizations.[1] But the weather data for Rwanda tells us that the heavy rain didn't start until most of those people had already died due to flooding and avalanches. This means it must have rained heavily on the Congolese side in the preceding days. But there are no weather data to show us exactly when and where. The area affected (around Lake Kivu) has a long history of conflict and instability influenced by colonialism; to this day, the region has major problems with the development of infrastructure and the service sector. In 2023, violent clashes between state and non-state groups led to large-scale displacement in South Kivu—both in the Democratic Republic of the Congo and in western Rwanda. The displaced live in harsh conditions in improvised emergency shelters, which makes them particularly susceptible to the consequences of extreme weather. The 2023 floods damaged the already-poor water, sanitation, and hygiene infrastructure and destroyed agricultural fields, heightening food insecurity and leading to outbreaks of diseases like cholera. Thousands of people fled the region.

How did it come to this? The clearing of forests for settlements, agriculture, and mining has led to severe soil erosion around Lake Kivu, increasing the risk of landslides. The less-than-effective government is so busy dealing with social conflicts that it barely has time to enforce regulations on construction and land use. In Congolese mines, rare materials are extracted for the global production of batteries and other components, particularly for renewable energies. But the forests—which are visibly disappearing—are an important sink for the carbon mostly produced by the burning of coal, oil, and gas in the Global North.

The region around Lake Kivu combines many of the factors that cause inequality and injustice. The effects of extreme weather events are amplified by mining, which contributes to the deterioration of land and water quality, promotes the abuse of workers, and stokes social conflicts. But as in most areas with long-standing conflicts, little data (on the weather and climate and on living conditions) is available for this region. My kind of research and science depends on data, which is why this book doesn't have a chapter dedicated to the climate, war, and conflict. If it did, it would be further evidence that climate change is less a physical problem and far more the result of the colonial-fossil narrative. The conditions in Congolese mines and their negative impact on the region clearly emphasize once again that stopping global warming is nowhere near enough.

Whether you're reading academic work or watching the daily bulletin, our warming world is an almost constant source of news about climate change. Whenever you're feeling gloomy, a news item or data point will be at hand to plunge you into existential crisis. At the same time, right-wing parties (which are anti-science and often have climate change denial as a key agenda point) are achieving record results across Europe,[2] and citizens of various European countries find the actions of protest groups like the Last Generation in Germany, or Just Stop Oil in the UK, to be exaggerated or wrong,[3] despite them only demanding that governments uphold the commitments they made in ratifying the Paris Agreement. And yet many Europeans and a consistently increasing number of Americans consider climate change to be one of the most important problems of the present day.[4] How does it all fit together?

Clearly, many people are afraid of climate change but don't want to do anything about it. As soon as activists point out this contradiction, they are met with incomprehension and even hate. European and Global North countries have the same two dominant narratives: "climate change is an unstoppable catastrophe that will soon destroy humanity" and "even the most cautious attempt to shift the status quo is an attack on personal freedom." Faced with these two irreconcilable beliefs, we bury our heads in the sand while clinging to the current state of affairs against all reason. But these two narratives have very little to do with the reality of climate change, and even less to do with what people actually need for a good life. Today we consider it unacceptable to go without a car, but this has nothing to do with our innermost needs and very little to do with convenience; it is the product of the automotive lobby, which has acted ruthlessly against the clamor of protests from people who would love more livable cities with no exhaust fumes or noise.[5] And numerous studies emphatically show that gross domestic product (GDP) is a totally unsuitable benchmark for quality of life. Way back in 1968, Robert Kennedy said that a country's GDP measures "everything... except that which makes life worthwhile."[6] But even today, decades later, politics is still being practiced as though the only thing that counts in a society is maximizing profits for a select few. Between 1930 and 1980, GDP grew and quality of life improved for everyone. Since then, quality of life has only improved for a small, rich demographic, while inequality increases in society as a whole—not just since the pandemic and Russia's invasion of Ukraine, but it has been particularly dramatic since then.[7]

It makes me angry that we are supposed to live our entire lives in the colonial-fossil narrative dictated by a handful of lobbies to maximize profits for a few and to the disadvantage of almost everyone else, particularly the poor, the less educated, people of color, and, above all, the people in the Global South. But mere anger is about as much use as knowing that we have been indoctrinated by industrial lobbies for decades. We don't lack knowledge. We know all about industry strategies by now; they've been tried and tested on tobacco, cars, pesticides, lead, coal, gas, and oil. And yet many journalists ask me, again and again, why we aren't doing anything about climate change. Some find my answer trite. To those for whom self-reflection is too much effort, it might even sound like a conspiracy theory. So what do we need to do if anything is actually going to change? In my opinion, we need to create new narratives and tell better stories. Narratives are incredibly powerful and are created by people themselves.

"Tipping points" are central to the public climate debate; these are parts of the climate system that change irreversibly under certain conditions when the global temperature rises beyond a certain threshold. But we can only estimate roughly the temperatures at which these processes occur. The actual tipping takes place over a period of decades or centuries. When exactly these tipping points will be exceeded is irrelevant for the measures we (would) need to take against climate change today. Even without these tipping points, the changes caused by greenhouse gas emissions are already so dramatic that there is no economic or ethical justification for not immediately putting a stop to fossil-fuel burning. Tipping points may be another argument for this, but they aren't crucial for real-world decisions. What they are good for is spreading fear. This is why the media like to talk about tipping points; fear sells, but it doesn't inspire people to act. All it does is paralyze them with shock. So talking about tipping points isn't useful when trying to replace the colonial-fossil narrative.

Instead, the focus should be on the lives and stories of Tema and Tamaria, Manuela and Jasmine, and Iracadju Ka'apor—who are among the people who have been quoted at the start of each chapter. In other words, the focus should be on how climate change affects specific people. This goes for both the media and the scientific community, which tends to sideline the real damage—the deaths, decimated crops, and wrecked houses—while concentrating excessively on the causes; national emissions inventories are taken regularly and standards are agreed on how to measure them. Let's compare our current situation to a crime scene: we know exactly who has bought and used which weapons and which ammunition was fired by whom. But the victims of the shootings aren't counted; in many cases they aren't even noted, and nobody bothers to secure the bullet casings or fingerprints as evidence. There'll be a brief media outcry after an especially bloody massacre, but nothing will change.

Climate change follows a similar pattern. There are no globally agreed standards for measuring or systematically recording damage. The evidence presented in this book is still composed of individual cases, and since there's no systematic overview of loss and damage, it's hugely difficult to compare pieces of evidence. This means that the damage caused by climate change can be exaggerated or downplayed however we wish. It's usually the latter—the available scientific evidence is usually viewed and analyzed through the colonial-fossil lens. Coming back to the crime analogy, the perpetrators might switch weapons now and again, but otherwise nobody bothers them, while the number of victims rises constantly. The lobbies representing the perpetrators tell us stories to distract from the corpses being carried away and make plausible how beneficial the use of the weapons is for humanity. Sounds absurd, doesn't it?

When we first encounter a new narrative, it usually sounds completely absurd. In February 2020, I couldn't have imagined my government prohibiting me from leaving the country. Just one month later, it became a natural government response to a global pandemic. In a matter of weeks, something we had long considered absurd became normality, and we could hardly imagine ever having seen it as absurd. New narratives often take longer to prevail. In the eighteenth century, the idea of women being persons in their own right was generally considered outrageous nonsense. It was an irrefutable fact that men were powerful and women inferior "by nature," and any scientific evidence to the contrary (either in hominids or in other animal species) was ignored. If nothing else, this is why we still have a completely misleading idea of how reproduction and evolution work; the reproductive organs, partner selection strategies, and brain chemistry of female animals were studied (if they were studied at all) with the unquestioned assumption that all female animals are passive and maternal. Spiders and bees were evolutionary anomalies by challenging the idea of the "strong male" and "passive female."

Science is slowly discovering how wrong this assumption is and all the things we have missed in 150 years of evolutionary biology due to our lack of insight into the female sex—with interesting findings on lemurs, hyenas, albatrosses, and geckos.[8] We'll probably have to wait a while longer for this information to find its way into nature documentaries. But the narrative is slowly changing. Today we tell ourselves the story that women have just as many rights and abilities as men, and that the idea of the gender binary isn't all that accurate. Of course, attempts are still being made to suppress the very idea of equality, but this is now only possible with brutal violence, as seen in Iran or Afghanistan.

Effective action will only be taken against climate injustice when we manage to tell better stories than the one about fear of tipping points, the one about never flying ever again, or the one about turning back globalization. We need constructive, powerful narratives that help to dismantle traditional, institutionalized, structural inequality, rather than reinforcing it through the consequences of climate change.

What would happen if we were to follow the example of historian and activist Rebecca Solnit and turn the story on its head that right now our lives are filled with hardships and deprivations and the riches and abundance are yet to come?[9] Whether you believe it or not, this is exactly what most people would get from meaningful measures to tackle climate change: a higher quality of life, better health, greater freedom to be who they want to be, better houses, more butterflies and bees... The only reason this seems absurd is because we have a downright distorted view of what deprivation and abundance mean in a positive sense. There's plenty of information out there on how much more livable a world would be with greener and more walkable cities; better-insulated houses; more freedom to walk, run, play, dance, and cycle; functioning social systems; and greater biodiversity—but we retaliate emotionally and perceive anything that deviates from the colonial-fossil narrative to be a sacrifice, an attack on our values, our culture, our traditions.

The word "we" is doing a lot of work here; hopefully, many of the people who read this are dreaming—just like I am—that all private vehicles will vanish overnight (replaced by a well-developed public transport network, of course). But we have all been socialized in this old narrative, and until we reflect on this, we won't realize how strongly it continues to affect us. In both Europe and North America, this is demonstrated not least by the aggressive reactions to climate activists.

If we want to change something, we need to do more than enlighten and inform; we need to tell better stories so powerful that they overwrite those told by the world's most influential corporations. This seems barely attainable because, as I have shown, it affects all levels of society. But what this book also shows are the various solutions being developed for these multifarious problems. So we can either paint catastrophic scenarios, panic, and give up, or we can realize that there are many areas that require improvement and we have plenty of agency to tackle them. Climate change isn't a reason to bury our heads in the sand; it's a reason to effect change together.

Right now "we" are not that many, and we have limited influence over society's powerful groups. It's frustrating to see how a majority of the demand for fossil fuels is generated by clever and lavishly funded industry campaigns encouraging people to keep burning gas and oil. The idea of flight shaming could ultimately have originated in the fossil-fuel industry because it furthers the idea that consumption is the only way an individual can affect anything, and that there is no freedom without fossil fuels.

In these circumstances, we can quickly feel powerless and find ourselves asking what an individual can do. The answer? Stop being an individual! This is one of the key ideas behind every form of societal change and every protest—when more and more people start to call for change, eventually it becomes inevitable. For twenty-eight years, East and West Germany were separated by a wall of deadly violence. What began as secret gatherings culminated in ever-larger protests until the pressure exerted by the people (alongside economic problems in the East German government and reforms in the Soviet Union) tore down not just the Berlin Wall, but a dictatorship's entire political system. This is a good model not just for joint efforts to combat climate change in general, but also for changing the underlying stories and narratives.

I wish I could present you with a clear plan for change. I try to show and explain how racism, colonialism, sexism, and climate change are connected to justice and how positive changes on one level can lead to positive effects on the others. To allow us to understand this properly and really imagine a better future, we need stories in newspapers and magazines, on television, in novels, picture books, theaters, and cinema, and also in the visual arts. But when I go to the Tate Modern, one of the most significant collections of contemporary art (and, miraculously, within walking distance of my home), I am excited by the exhibitions and yet a little disappointed that hardly any of them tackle climate change, and the few that do don't fulfill art's true potential. Art is beautiful and evocative and thrilling and unsettling, and this is what makes it necessary. Art is the opposite of deprivation, the thing we flee to to process trauma, but also the place we go to to gain strength and understand more about humanity. Thanks to Maya Angelou, we all know why the caged bird sings. Because of Toni Morrison, the brutal realities of American slavery and their ongoing legacy have been memorialized for generations of readers. We also need art to find ideal worlds and heroines to create them. If art and pop culture have shown again and again the power of questioning the status quo, who is giving us new visions to aspire to? What hero is going to show us how to question the colonial-fossil narrative, in a moment when that challenge is needed now more than ever?

Acknowledgments

I N THE TWO YEARS it took to write this book, everything in my life changed. A new city, new job, new school, new house, new dog, new cat. But all the truly important things are still there even if it wasn't always easy and I couldn't take them for granted. Thank you, Alek. Thank you, Hanni. Thank you, Matt.

I also thank my Mama and my little sister.

Life, books, and science are always a collective effort, and the risk of leaving someone out is far too great. You know who you are. Thank you, team!

Notes

CHAPTER I: INEQUALITY IN THE SPOTLIGHT

1. Tim Cowan, Sabine Undorf, Gabriele C. Hegerl, Luke J. Harrington, and Friederike E. L. Otto, "Present-day greenhouse gases could cause more frequent and longer Dust Bowl heat waves," *Nature Climate Change* 10 (2020): 505–510, doi.org/10.1038/s41558-020-0771-7

2. R. F. Stuart-Smith, G. H. Roe, S. Li, and M. R. Allen, "Increased outburst flood hazard from Lake Palcacocha due to human-induced glacier retreat," *Nature Geoscience* 14 (2021): 85–90, doi.org/10.1038/s41561-021-00686-4

3. unfccc.int/process-and-meetings/the-paris-agreement

4. See the explanations of "hazard" on the UNDRR website: undrr.org/terminology/hazard

5. Mariam Zachariah, Clair Barnes, Friederike E. L. Otto, et al, "Climate change exacerbated heavy rainfall leading to large scale flooding in highly vulnerable communities in West Africa," World Weather Attribution (2022), worldweatherattribution.org/wp-content/uploads/Nigeriafloods_scientific-report.pdf

6. Ibid.

7. Maria Fernandes-Jesus, Brendon Barnes, and Raquel F. Diniz, "Communities reclaiming power and social justice in the face of climate change," *Community Psychology in Global Perspective (CPGP)* 6, no. 2/2 (2020): 1–21, doi.org/10.1285/i24212113v6i2-2p1

8. Reto Knutti, Joeri Rogelj, Jan Sedláček, and Erich M. Fischer, "A scientific critique of the two-degree climate change target," *Nature Geoscience* 9 (2016): 13–18, doi.org/10.1038/ngeo2595

9. Alex Epstein, "How fossil fuels cleaned up our environment," *Forbes*, January 28, 2015, forbes.com/sites/alexepstein/2015/01/28/how-fossil-fuels-cleaned-up-our-environment

CHAPTER 2: A CONTINENT OFF THE CHARTS

1. "Canada: Disastrous impact of extreme heat," Human Rights Watch, October 5, 2021, hrw.org/news/2021/10/05/canada-disastrous-impact-extreme-heat

2. "Physicians, lawyers call on BC to investigate thousands of heat dome injuries," West Coast Environmental Law, July 29, 2021, wcel.org/media-release/physicians-lawyers-call-bc-investigate-thousands-heat-dome-injuries

3. Canada: Disastrous impact of extreme heat," Human Rights Watch, October 5, 2021, hrw.org/news/2021/10/05/canada-disastrous-impact-extreme-heat

4. See tourist information on Lytton: britishcolumbia.com/plan-your-trip/regions-and-towns/vancouver-coast-mountains/lytton

5. "New evacuation orders issued in western Canada as fire guts town after record heat," *Reuters*, July 2, 2021, reuters.com/business/environment/wildfire-forces-evacuation-residents-small-western-canada-town-2021-07-01

6. Ibid.

7. "Western North American extreme heat virtually impossible without human-caused climate change," World Weather Attribution, July 7, 2021, worldweatherattribution.org/western-north-american-extreme-heat-virtually-impossible-without-human-caused-climate-change

8. Sjoukje Y. Philip, Sarah F. Kew, Friederike E. L. Otto, et al., "Rapid attribution analysis of the extraordinary heat wave on the Pacific coast of the US and Canada in June 2021," *Earth System Dynamics* 13, no. 4 (2022), doi.org/10.5194/esd-13-1689-2022

9. David Wallace-Wells, "Can you even call deadly heat 'extreme' anymore?," *New York Times*, May 17, 2022, nytimes.com/2022/05/17/opinion/india-heat-wave-pakistan-climate-change.html

10. See chapter 11 of the IPCC Sixth Assessment Report: ipcc.ch/report/ar6/wg1/downloads/report/IPCC_AR6_WGI_Chapter11.pdf

11. Ibid.

12. For India and Pakistan: "Climate change made devastating early heat in India and Pakistan 30 times more likely," World Weather Attribution, May 23, 2022, worldweatherattribution.org/climate-change-made-devastating-early-heat-in-india-and-pakistan-30-times-more-likely

    For Argentina: "Climate change made record breaking early season heat in Argentina and Paraguay about 60 times more likely," World Weather Attribution, December 21, 2022, worldweatherattribution.org/climate-change-made-record-breaking-early-season-heat-in-argentina-and-paraguay-about-60-times-more-likely

13. See chapter 11 of the IPCC Sixth Assessment Report: ipcc.ch/report/ar6/wg1/downloads/report/IPCC_AR6_WGI_Chapter11.pdf

14. James E. Overland, "Causes of the record-breaking Pacific Northwest heatwave, late June 2021," *Atmosphere* 12, no. 11 (2021), doi.org/10.3390/atmos12111434

   Chunzai Wang, Jiayu Zheng, Wei Lin, and Yuqing Wang, "Unprecedented heatwave in western North America during late June of 2021: Roles of atmospheric circulation and global warming," *Advances in Atmospheric Sciences* 40 (2023): 14–28, doi.org/10.1007/s00376-022-2078-2

15. "Physicians, lawyers call on BC to investigate thousands of heat dome injuries," West Coast Environmental Law, July 29, 2021, wcel.org/media-release/physicians-lawyers-call-bc-investigate-thousands-heat-dome-injuries

16. Luke J. Harrington, Kristie L. Ebi, David J. Frame, and Friederike E. L. Otto, "Integrating attribution with adaptation for unprecedented future heatwaves," *Climatic Change* 172 (2022), doi.org/10.1007/s10584-022-03357-4

17. Sjoukje Y. Philip, Sarah F. Kew, Friederike E. L. Otto, et al., "Rapid attribution analysis of the extraordinary heat wave on the Pacific coast of the US and Canada in June 2021," *Earth System Dynamics* 13, no. 4 (2022): 1689–1713, doi.org/10.5194/esd-13-1689-2022

18. Ambarish Vaidyanathan, Josephine Malilay, Paul Schramm, and Shubhayu Saha, "Heat-related deaths—United States, 2004–2018," *Morbidity and Mortality Weekly Report (MMWR)* 69, no. 24 (2020): 729–734, dx.doi.org/10.15585/mmwr.mm6924a1

19. Kate R. Weinberger, Daniel Harris, Keith R. Spangler, et al., "Estimating the number of excess deaths attributable to heat in 297 United States counties," *Environmental Epidemiology* 4, no. 3 (2020): e096, doi.org/10.1097/EE9.0000000000000096

20. "Heat mortality monitoring report: 2020," gov.uk/government/publications/heat-mortality-monitoring-reports/heat mortality-monitoring-report-2020

   "Extreme heat and human mortality: A review of heat-related deaths in B.C. in summer 2021," Report to the Chief Coroner of British Columbia (2022), gov.bc.ca/assets/gov/birth-adoption-death-marriage-and-divorce/deaths/coroners-service/death-review-panel/extreme_heat_death_review_panel_report.pdf

21. "Number of deaths due to hurricanes in the United States from 2000 to 2021," Statista (2022), statista.com/statistics/203729/fatalities-caused-by-tropical-cyclones-in-the-us

22. Kim Stanley Robinson, *The Ministry for the Future* (Orbit Books, 2020)

23. "Canada: Disastrous impact of extreme heat," Human Rights Watch (2021), hrw.org/news/2021/10/05/canada-disastrous-impact-extreme-heat

24. Kathryn Blaze Baum and Matthew McClearn, "Extreme, deadly heat in Canada is going to come back, worse than ever. Will we be ready?," *The Globe and Mail*, September 25, 2021, theglobeandmail.com/canada/article-extreme-deadly-heat-in-canada-is-going-to-come-back-and-worse-will-we

25. Sjoukje Y. Philip, Sarah F. Kew, Friederike E. L. Otto, et al., "Rapid attribution analysis of the extraordinary heatwave on the Pacific Coast of the US and Canada June 2021," World Weather Attribution, worldweatherattribution.org/wp-content/uploads/NW-US-extreme-heat-2021-scientific-report-WWA.pdf

26. Gene Johnson and Sarah Cline, "Northwest US faces hottest day of intense heat wave," *The Seattle Times*, June 28, 2021, seattletimes.com/seattle-news/oregon-washington-idaho-hang-on-for-hottest-day-of-intense-heat-wave

27. Matthew W. Jones, Adam Smith, Richard Betts, et al., "Climate change increases the risk of wildfires," *ScienceBrief* (2020), preventionweb.net/files/73797_wildfiresbriefingnote.pdf

28. Kathryn Blaze Baum and Matthew McClearn, "Extreme, deadly heat in Canada is going to come back, worse than ever. Will we be ready?," *The Globe and Mail*, September 25, 2021, theglobeandmail.com/canada/article-extreme-deadly-heat-in-canada-is-going-to-come-back-and-worse-will-we

29. Friederike Otto, *Angry Weather* (Greystone Books, 2020), 161–181

30. See, for example, the recommendations for action for drawing up heat action plans from the German Federal Ministry of Health: bundesgesundheitsministerium.de/service/begriffe-von-a-z/h/hitze-hitzeaktionsplaene.html

31. Jonas Schwaab, Ronny Meier, Gianluca Mussetti, et al., "The role of urban trees in reducing land surface temperatures in European cities," *Nature Communications* 12 (2021), doi.org/10.1038/s41467-021-26768-w

32. Richard Fuller, Philip J. Landrigan, Kalpana Balakrishnan, et al., "Pollution and health: A progress update," *The Lancet Planetary Health* 6, no. 6 (2022): 535–545, doi.org/10.1016/S2542-5196(22)00090-0

33. Yuval Noah Harari, *Sapiens: A Brief History of Humankind* (Harper, 2015)

34. Press release by EY, an organization belonging to Ernst & Young Global Limited, December 29, 2022: ey.com/en_ch/news/2022-press-releases/12/us-companies-are-dominating-stock-exchanges-globally

CHAPTER 3: AN AFRICAN PHANTOM?

1. Simon S. Cordall, "'We can't endure this': Migrants suffer in extreme Tunisian heat," *Al Jazeera*, July 24, 2023, aljazeera.com/news/2023/7/24/we-cant-endure-this-migrants-suffer-in-extreme-tunisian-heat

2. "Protecting people from extreme heat," Human Rights Watch, July 21, 2022, hrw.org/news/2022/07/21/protecting-people-extreme-heat

3. Albert E. Manyuchi, Coleen Vogel, Caradee Y. Wright, and Barend Erasmus, "The self-reported human health effects associated with heat exposure in Agincourt sub-district of South Africa," *Humanities and Social Sciences Communications* 9 (2022), doi.org/10.1057/s41599-022-01063-1

4. Shantelle Spencer, Tida Samateh, Katharina Wabnitz, et al., "The challenges of working in the heat whilst pregnant: Insights from Gambian women farmers in the face of climate change," *Frontiers in Public Health* 10 (2022), doi.org/10.3389/fpubh.2022.785254

5. See description on the EM-DAT website: emdat.be

6. S. E. Perkins-Kirkpatrick and S. C. Lewis, "Increasing trends in regional heatwaves," *Nature Communications* 11 (2020), doi.org/10.1038/s41467-020-16970-7

7. Camilo Mora, Bénédicte Dousset, Iain Caldwell, et al., "Global risk of deadly heat," *Nature Climate Change* 7 (2017): 501–506, doi.org/10.1038/nclimate3322

8. Colin Raymond, Tom Matthews, and Radley M. Horton, "The emergence of heat and humidity too severe for human tolerance," *Science Advances* 6, no. 19 (2020), doi.org/10.1126/sciadv.aaw1838

9. Jakob Zscheischler, Olivia Martius, Seth Westra, et al., "A typology of compound weather and climate events," *Nature Reviews Earth & Environment* 1 (2020): 333–347, doi.org/10.1038/s43017-020-0060-z

10. "Human-induced climate change increased drought severity in Horn of Africa," World Weather Attribution, April 27, 2023, worldweatherattribution.org/human-induced-climate-change-increased-drought-severity-in-southern-horn-of-africa

11. NASA Fire Information for Resource Management System: firms.modaps.eosdis.nasa.gov

12. Matthew W. Jones, John T. Abatzoglou, Sander Veraverbeke, et al., "Global and regional trends and drivers of fire under climate change," *Reviews of Geophysics* 60, no. 3 (2022), doi.org/10.1029/2020RG000726

13. Historic data on air quality is available at iqair.com

14. Adenike Oladosu's speech at the Schauspiel Stuttgart in 2021: environewsnigeria.com/climate-change-nowhere-is-safe-if-africa-isnt-adenike-oladosu-warns

15. Rajashree Kotharkar and Aveek Ghosh, "Progress in extreme heat management and warning systems: A systematic review of heat-health action plans (1995–2020)," *Sustainable Cities and Society* 76 (2022), doi.org/10.1016/j.scs.2021.103487

16. Jeremy J. Hess, Sathish LM, Kim Knowlton, et al., "Building resilience to climate change: Pilot evaluation of the impact of India's first heat action plan on all-cause mortality," *Journal of Environmental and Public Health* (2018), doi.org/10.1155/2018/7973519

17. Mireille A. Folkerts, Peter Bröde, W. J. Wouter Botzen, et al., "Long term adaptation to heat stress: Shifts in the minimum mortality temperature in the Netherlands," *Frontiers in Physiology* 11 (2020), doi.org/10.3389/fphys.2020.00225

18. See chapter 8 of the IPCC Sixth Assessment Report, on poverty, livelihoods, and sustainable development: ipcc.ch/report/ar6/wg2/downloads/report/IPCC_AR6_WGII_Chapter08.pdf

19. Shantelle Spencer, Tida Samateh, Katharina Wabnitz, et al., "The challenges of working in the heat whilst pregnant: Insights from Gambian women farmers in the face of climate change," *Frontiers in Public Health* 10 (2022), doi.org/10.3389/fpubh.2022.785254

20. Joan Flocks, Valerie Vi Thien, Jennifer Runkle, et al., "Female farmworkers' perceptions of heat-related illness and pregnancy health," *Journal of Agromedicine* 18, no. 4 (2013): 350–358, doi.org/10.1080/1059924X.2013.826607

21. See information on The Gambia's agricultural industry on the U.S. International Trade Administration website: trade.gov/country-commercial-guides/gambia-agriculture

22. See chapter 8 of the IPCC Sixth Assessment Report by Working Group II, on poverty, livelihoods, and sustainable development: 1192, ipcc.ch/report/ar6/wg2/downloads/report/IPCC_AR6_WGII_Chapter08.pdf

Also see Goal Five of UN Department of Economic and Social Affairs: sdgs.un.org/goals

CHAPTER 4: WHEN JUSTICE DRIES UP

1. See: Daisy Dunne, "Mapped: African world heritage sites threatened by sea level rise 'to triple by 2050,'" *Carbon Brief*, February 10, 2022, carbonbrief.org/mapped-african-world-heritage-sites-threatened-by-sea-level-rise-to-triple-by-2050

2. Wendell Roelf, "Drought forces some Cape Town residents into midnight queues for water," *Reuters*, February 6, 2018, reuters.com/article/us-safrica-drought-lines/drought-forces-some-cape-town-residents-into-midnight-queues-for-water-idUSKBN1FQ26M

3. Andrew Harding, "Cape Town's Day Zero: 'We are axing trees to save water,'" *BBC News*, November 9, 2021, bbc.co.uk/news/world-africa-59221823

4. Jon Heggie, "Day Zero: Where next?," *National Geographic*, nationalgeographic. com/science/article/partner-content-south-africa-danger-of-running-out-of-water

5. On the consumption of drinking water in the EU and U.K., see: europarl.europa. eu/news/en/headlines/society/20181011STO15887/drinking-water-in-the-eu-better-quality-and-access

6. Robyn Dixon, "Global development: How Cape Town found water savings California never dreamed of," *Los Angeles Times*, April 1, 2018, latimes.com/ world/africa/la-fg-south-africa-drought-20180401-story.html

7. Western Cape Government website and its up-to-date information on dam levels: westerncape.gov.za/general-publication/latest-western-cape-dam-levels

8. "Climate change-exacerbated rainfall causing devastating flooding in Eastern South Africa," World Weather Attribution, May 13, 2022, worldweatherattribution.org/climate-change-exacerbated-rainfall-causing-devastating-flooding-in-eastern-south-africa

9. Jon Heggie, "Day Zero: Where next?," *National Geographic*, nationalgeographic. com/science/article/partner-content-south-africa-danger-of-running-out-of-water

10. Jackie Dugard, "Water rights in a time of fragility: An exploration of contestation and discourse around Cape Town's 'Day Zero' water crisis," *Water* 13, no. 22 (2021), doi.org/10.3390/w13223247

11. See chapter 11 (by Working Group I) of the IPCC Sixth Assessment Report (AR6-WGI): 1557 et seq., ipcc.ch/report/ar6/wgi/downloads/report/IPCC_AR6_WGI_Chapter11.pdf

12. Ibid.

13. Chapter 4 of the report by Working Group II, IPCC Sixth Assessment Report (AR6-WGII). The full English-language report can be found at: ipcc.ch/report/ ar6/wg2. On the consequences of agricultural droughts, see chapter 4, pages 578 et seq., particularly page 584.

14. Friederike E. L. Otto, Piotr Wolski, Flavio Lehner, et al., "Anthropogenic influence on the drivers of the Western Cape drought 2015–2017," *Environmental Research Letters* 13, no. 12 (2018), doi.org/10.1088/1748-9326/aae9f9

15. Ibid.

16. Official Western Cape agricultural profile for 2021; see figure 4.1, elsenburg. com/wp-content/uploads/2022/07/2021-WC-Agric-Sector-Profile-1.pdf

17. Official description of Cape Town's water supply, resource.capetown.gov.za/ documentcentre/Documents/Graphics%20and%20educational%20material/ Water%20Services%20and%20Urban%20Water%20Cycle.pdf

18. Emma Luker and Leila M. Harris, "Developing new urban water supplies: Investigating motivations and barriers to groundwater use in Cape Town," *International Journal of Water Resources Development* 35, no. 6 (2019): 917–937, doi. org/10.1080/07900627.2018.1509787

19. The official water outlook for the Western Cape in 2018: greencape.co.za/ assets/Uploads/Water-Outlook-2018-Rev-22-updated-23-March-2019.pdf

20. Description on the "2030 Water Resources Group" platform founded by the World Bank, 2030wrg.org/where-we-work/sao-paulo-brazil

21. "Economic review and outlook 2017": 49, westerncape.gov.za/assets/ departments/treasury/Documents/Research-and-Report/2017/2017_pero_ printers_proof_21_september_2017_f.pdf

22. Mike Muller, "Lessons from Cape Town's drought," *Nature* 559 (2018): 174–176, media.nature.com/original/magazine-assets/d41586-018-05649-1/d41586-018-05649-1.pdf

23. Mike Muller, "Understanding the origins of Cape Town's water crisis," *Civil Engineering* (2017): 11–16, papers.ssrn.com/sol3/papers.cfm?abstract_id= 2995937

24. Mike Muller, "Lessons from Cape Town's drought," *Nature* 559 (2018): 175, media.nature.com/original/magazine-assets/d41586-018-05649-1/d41586-018-05649-1.pdf

25. Remy Caball and Shirin Malekpour, "Decision making under crisis: Lessons from the Millennium Drought in Australia," *International Journal of Disaster Risk Reduction* 34 (2019): 387–396, doi.org/10.1016/j.ijdrr.2018.12.008

26. Vanessa Lucena Empinotti, Jessica Budds, and Marcelo Aversa, "Governance and water security: The role of the water institutional framework in the 2013–15 water crisis in São Paulo, Brazil," *Geoforum* 98 (2019): 46–54, doi.org/10.1016/j. geoforum.2018.09.022

27. Jonathan Watts, "Brazil's worst drought in history prompts protests and blackouts," *The Guardian*, January 23, 2015, theguardian.com/world/2015/jan/23/ brazil-worst-drought-history

28. Drought policy brief by the Bureau for Food and Agricultural Policy (BFAP): bfap. co.za/wp-content/uploads/2023/05/DroughtPolicyBrief_2018.pdf

29. M. Fanadzo, B. Ncube, A. French, and A. Belete, "Smallholder farmer coping and adaptation strategies during the 2015–18 drought in the Western Cape, South Africa," *Physics and Chemistry of the Earth* 124, no. 1 (2021), doi.org/ 10.1016/j.pce.2021.102986

30. Jasper Verschuur, Sihan Li, Piotr Wolski, and Friederike E. L. Otto, "Climate change as a driver of food insecurity in the 2007 Lesotho-South Africa drought," *Scientific Reports* 11, 3852 (2021), doi.org/10.1038/s41598-021-83375-x

31. Joint statement by the heads of the Food and Agriculture Organization, International Monetary Fund, World Bank Group, World Food Programme and World Trade Organization on the global food and nutrition security crisis, worldbank.org/en/news/statement/2023/02/08/joint-statement-on-the-global-food-and-nutrition-security-crisis

32. "Climate change made devastating early heat in India and Pakistan 30 times more likely," World Weather Attribution, May 23, 2022, worldweatherattribution.org/climate-change-made-devastating-early-heat-in-india-and-pakistan-30-times-more-likely

33. David J. Frame, Michael F. Wehner, Ilan Noy, and Suzanne M. Rosier, "The economic costs of Hurricane Harvey attributable to climate change," *Climatic Change* 160 (2020): 271–281, doi.org/10.1007/s10584-020-02692-8

34. James Rising, Marco Tedesco, Franziska Piontek, and David A. Stainforth, "The missing risks of climate change," *Nature* 610 (2022): 643–651, doi.org/10.1038/s41586-022-05243-6

35. Jasper Verschuur, Sihan Li, Piotr Wolski, and Friederike E. L. Otto, "Climate change as a driver of food insecurity in the 2007 Lesotho-South Africa drought," *Scientific Reports* 11, 3852 (2021), doi.org/10.1038/s41598-021-83375-x

36. HDI data can be viewed in detail and downloaded on the HDI website: hdr.undp.org/data-center/human-development-index#/indicies/HDI

37. Data on Lesotho published by the World Bank: worldbank.org/en/country/lesotho/overview#1

38. See: gov.ls/economy

39. Jasper Verschuur, Friederike E. L. Otto, and Piotr Wolski, "How climate change drove food insecurity in the 2007 Lesotho drought," *Carbon Brief* (2021), carbonbrief.org/guest-post-how-climate-change-drove-food-insecurity-in-the-2007-lesotho-drought

40. Robert Belano, "Day Zero is already here for Cape Town's poor," *Left Voice*, March 30, 2018, leftvoice.org/day-zero-is-already-here-for-cape-town-s-poor

41. Aryn Baker, "What it's like to live through Cape Town's massive water crisis," *Time* (2023), time.com/cape-town-south-africa-water-crisis

42. Robert Belano, "Day Zero is already here for Cape Town's poor," *Left Voice*, March 30, 2018, leftvoice.org/day-zero-is-already-here-for-cape-town-s-poor

43. Natalie McCauley, "Water rights and Day Zero: Perspectives on the Cape Town water crisis" (2019), law.ucla.edu/sites/default/files/PDFs/Academics/McCauley-Water%20Rights%20and%20Day%20Zero.pdf

44. Ivan Turok, Andreas Scheba, Justin Visagie, "Can social housing help to integrate divided cities?," *Cambridge Journal of Regions, Economy and Society* 15, no. 1 (2022): 93–116, doi.org/10.1093/cjres/rsab031

CHAPTER 5: POVERTY: THE ROOT OF THE CRISIS

1. "Don't look the other way: Madagascar in the grip of drought and famine," *ReliefWeb*, July 9, 2021, reliefweb.int/report/madagascar/don-t-look-other-way-madagascar-grip-drought-and-famine

2. Food Security Information Network, "Global report on food crises 2023": 107–108, fsinplatform.org/sites/default/files/resources/files/GRFC2023-country-madagascar.pdf

3. United Nations Development Programme Human Development Index: hdr. undp.org/data-center/specific-country-data#/countries/MDG

4. Ghislain Vieilledent, Clovis Grinand, Fety A. Rakotomalala, et al., "Combining global tree cover loss data with historical national forest cover maps to look at six decades of deforestation and forest fragmentation in Madagascar," *Biological Conservation* 222 (2018): fig. 2, doi.org/10.1016/j.biocon.2018.04.008

5. "WFP: Madagascar families facing world's first potential climate change famine," *UN News*, October 21, 2021, news.un.org/en/audio/2021/10/1103682

6. Luke J. Harrington, Piotr Wolski, Izidine Pinto, et al., "Limited role of climate change in extreme low rainfall associated with southern Madagascar food insecurity, 2019–21," *Environmental Research: Climate* 1, no. 2 (2022), doi.org/10.1088/2752-5295/aca695

7. Frank-Borge Wietzke, "Institution building from the bottom and the top: Long-term consequences of missionary activity and settler colonialism in Madagascar," October 18, 2018, dx.doi.org/10.2139/ssrn.3255653

8. Ibid.

9. Emmanuel Raju, Emily Boyd, and Friederike Otto, "Stop blaming the climate for disasters," *Communications Earth & Environment* 3, no. 1 (2022), doi.org/10.1038/s43247-021-00332-2

10. Marina Hülssiep, Thomas Thaler, and Sven Fuchs, "The impact of humanitarian assistance on post-disaster social vulnerabilities: Some early reflections on the Nepal earthquake in 2015," *Disasters* 45, no. 3 (2020), doi.org/10.1111/disa.12437

11. Ibid.

12. Paolo Cianconi, Sophia Betrò, and Luigi Janiri, "The impact of climate change on mental health: A systematic descriptive review," *Frontiers in Psychiatry* 11 (2020), doi.org/10.3389/fpsyt.2020.00074

13. K. B. Griffin and J. L. Enos, "Foreign assistance: Objectives and consequences," *Economic Development and Cultural Change* 18, no. 3 (1970): 313–327, jstor.org/stable/1152061

14. Thomas P. Higginbottom, Roshan Adhikari, and Timothy Foster, "Rapid expansion of irrigated agriculture in the Senegal River Valley following the 2008 food

price crisis," *Environmental Research Letters* 18 (2023), doi.org/10.1088/1748-9326/acaa46

15. Lazenya Weekes-Richemond, "Dear white women in international development," *Amplify*, August 18, 2021, medium.com/amplify/dear-white-women-in-international-development-4164f5f219a0

16. S. E. Perkins-Kirkpatrick and S. C. Lewis, "Increasing trends in regional heat waves," *Nature Communications* 11 (2020), doi.org/10.1038/s41467-020-16970-7

17. "Climate change increased rainfall associated with tropical cyclones hitting highly vulnerable communities in Madagascar, Mozambique & Malawi," World Weather Attribution, April 11, 2022, worldweatherattribution.org/climate-change-increased-rainfall-associated-with-tropical-cyclones-hitting-highly-vulnerable-communities-in-madagascar-mozambique-malawi

18. "Smart sovereignty," *DGAP Report* 1 (2022), dgap.org/sites/default/files/article_pdfs/smart-sovereignty_ideenwerkstatt-aussenpolitik_en-final.pdf

19. Vally Koubi, "Climate change and conflict," *Annual Review of Political Science* 22 (2019): 343–360, doi.org/10.1146/annurev-polisci-050317-070830

20. Lisa Thalheimer, David S. Williams, Kees van der Geest, and Friederike Otto, "Advancing the evidence base of future warming impacts on human mobility in African drylands," *Earth's Future* 9, no. 10 (2021), doi.org/10.1029/2020EF001958

CHAPTER 6: THE END OF THE RAINFOREST

1. "Rainforest Mafias: How violence and impunity fuel deforestation in Brazil's Amazon," Human Rights Watch, September 17, 2019, hrw.org/report/2019/09/17/rainforest-mafias/how-violence-and-impunity-fuel-deforestation-brazils-amazon

2. Catholic News Service, "Pope Francis urges action to save burning Amazon rainforest," *America*, August 26, 2019, americamagazine.org/faith/2019/08/26/pope-francis-urges-action-save-burning-amazon-rainforest

3. "Rainforest Mafias: How violence and impunity fuel deforestation in Brazil's Amazon," Human Rights Watch, September 17, 2019, hrw.org/report/2019/09/17/rainforest-mafias/how-violence-and-impunity-fuel-deforestation-brazils-amazon

4. Cecelia Smith-Schoenwalder, "Amazon rainforest burning in record fires," *U.S. News*, August 21, 2019, usnews.com/news/world-report/articles/2019-08-21/amazon-rainforest-burning-in-record-fires

5. NASA, "Reflecting on a tumultuous Amazon fire season," *NASA Earth Observatory*, March 4, 2020, earthobservatory.nasa.gov/images/146355/reflecting-on-a-tumultuous-amazon-fire-season

6. Ibid.

7. Ibid.

8. See NASA FIRMS fire map: firms.modaps.eosdis.nasa.gov/map

9. NASA, "Reflecting on a tumultuous Amazon fire season," *NASA Earth Observatory*, March 4, 2020, earthobservatory.nasa.gov/images/146355/ reflecting-on-a-tumultuous-amazon-fire-season

10. Global Forest Watch, "Global primary forest loss," gfw.global/3tPD5IY

11. Marcondes G. Coelho-Junior, Ana P. Valdiones, Julia Z. Shimbo, et al., "Unmasking the impunity of illegal deforestation in the Brazilian Amazon: A call for enforcement and accountability," *Environmental Research Letters* 17, no. 4 (2022), doi.org/10.1088/1748-9326/ac5193

12. "Rainforest Mafias: How violence and impunity fuel deforestation in Brazil's Amazon," Human Rights Watch, September 17, 2019, hrw.org/report/2019/ 09/17/rainforest-mafias/how-violence-and-impunity-fuel-deforestation-brazils-amazon

13. Hannah Ritchie provides a good overview of the status and consequences of deforestation in "Deforestation and forest loss" (February 4, 2021): ourworldindata.org/deforestation

14. See: ipcc.ch/sr15/chapter/spm

15. R. F. Stuart-Smith, B. J. Clarke, L. J. Harrington, and Friederike E. L. Otto, "Global climate change impacts attributable to deforestation driven by the Bolsonaro administration: Expert report for submission to the International Criminal Court" (2021), smithschool.ox.ac.uk/sites/default/files/2022-03/ICC_ report_final-sept-2021.pdf

16. "Deforestation in Brazil's Amazon falls 66% in August," *Reuters*, September 5, 2023, reuters.com/business/environment/deforestation-brazils-amazon-falls-70-august-2023-09-05

17. Friederike E. L. Otto and Emmanuel Raju, "Harbingers of decades of unnatural disasters," *Communications Earth & Environment* 4 (2023), doi.org/10.1038/ s43247-023-00943-x

18. R. F. Stuart-Smith, B. J. Clarke, L. J. Harrington, and Friederike E. L. Otto, "Global climate change impacts attributable to deforestation driven by the Bolsonaro administration: Expert report for submission to the International Criminal Court" (2021), smithschool.ox.ac.uk/sites/default/files/2022-03/ICC_ report_final-sept-2021.pdf

19. "Vulnerability and high temperatures exacerbate impacts of ongoing drought in Central South America," World Weather Attribution, February 16, 2023,

worldweatherattribution.org/vulnerability-and-high-temperatures-exacerbate-impacts-of-ongoing-drought-in-central-south-america

20. Chapter 2 of the Synthesis Report of the IPCC Sixth Assessment Report, "Current status and trends," report.ipcc.ch/ar6syr/pdf/IPCC_AR6_SYR_Longer-Report.pdf

21. Li Shilan et al., "Anthropogenic climate change contribution to wildfire-prone weather conditions in the Cerrado and arc of deforestation," *Environmental Research Letters* 16, no. 9 (2021), doi.org/10.1088/1748-9326/ac1e3a

22. "Safe climate: A report of the Special Rapporteur on Human Rights and the Environment," United Nations Environment Programme, October 2, 2019: 34, unep.org/resources/report/safe-climate-report-special-rapporteur-human-rights-and-environment

23. Peter Yeung, "A Peruvian farmer battles a German fossil fuel giant," *Deutsche Welle*, November 13, 2023, dw.com/en/a-peruvian-farmer-battles-a-german-fossil-fuel-giant/a-67019583

24. Climate Change Litigation Databases: climatecasechart.com

25. Climate Change Litigation Databases, "PSB et al. v. Brazil (on Climate Fund)" (2020), climate-laws.org/geographies/brazil/litigation_cases/psb-et-al-v-brazil-on-climate-fund

26. Ibid.

27. See the Paris Agreement (2015): unfccc.int/files/essential_background/convention/application/pdf/english_paris_agreement.pdf

28. Thomas Fischermann, "'Diese Regierung besteht aus Knallköpfen,'" *Zeit Online*, September 22, 2022, zeit.de/2022/39/amazonas-zerstoerung-regenwald-brasilien-politik/komplettansicht

29. Rome Statute of the International Criminal Court (1998): icc-cpi.int/sites/default/files/RS-Eng.pdf

30. Climate Change Litigation Databases, "The Planet v. Bolsonaro" (2021), climate-laws.org/geographies/international/litigation_cases/the-planet-v-bolsonaro

31. Friederike E. L. Otto, Petra Minnerop, Emmanuel Raju, et al., "Causality and the fate of climate litigation: The role of the social superstructure narrative," *Global Policy* 13, no. 5 (2022), doi.org/10.1111/1758-5899.13113

CHAPTER 7: FROM PAWN TO GAME CHANGER

1. Stephanie Gardiner, "Bushfires test Australia's volunteer firefighters to their limits," *The Guardian*, December 15, 2019, theguardian.com/australia-news/2019/dec/15/bushfires-australia-shattered-volunteer-firefighters-tested-to-their-limits

2. Lisa Cox and Nick Evershed, "'It's heart-wrenching': 80% of Blue Mountains and 50% of Gondwana rainforests burn in bushfires," *The Guardian*, January 16, 2020, theguardian.com/environment/2020/jan/17/its-heart-wrenching-80-of-blue-mountains-and-50-of-gondwana-rainforests-burn-in-bushfires

3. "Digital news report: Australia 2020," News & Media Research Centre, University of Canberra, apo.org.au/sites/default/files/resource-files/2020-06/apo-nid305057_0.pdf

4. Karl Mathiesen, "Brian Cox's sceptic takedown: A low point for climate journalism?," *Climate Home News*, August 18, 2016, climatechangenews.com/2016/08/18/brian-coxs-sceptic-takedown-a-low-point-for-climate-journalism

5. Christopher Wright, Daniel Nyberg, and Vanessa Bowden, "Beyond the discourse of denial: The reproduction of fossil fuel hegemony in Australia," *Energy Research & Social Science* 77 (2021), doi.org/10.1016/j.erss.2021.102094

6. NewGenCoal, "Climate change is a real problem. NewGenCoal is all about solutions," YouTube video, 3:00, April 27, 2010, youtube.com/watch?v=lyEt3lGQVWw

7. Ralph Hillman, "Fuel switch not long-term solution," *The Australian*, March 29, 2008: 5. Also see Christopher Wright, Daniel Nyberg, and Vanessa Bowden, "Beyond the discourse of denial: The reproduction of fossil fuel hegemony in Australia," *Energy Research & Social Science* 77 (2021), doi.org/10.1016/j.erss.2021.102094

8. Richard Conniff, "The myth of clean coal," *Yale Environment 360*, June 2, 2008, e360.yale.edu/features/the_myth_of_clean_coal

9. Jia C. Liu, Gavin Pereira, Sarah A. Uhl, et al., "A systematic review of the physical health impacts from non-occupational exposure to wildfire smoke," *Environmental Research* 136 (2015): 120–132, doi.org/10.1016/j.envres.2014.10.015

10. "What were climate scientists predicting in the 1970s?," *Skeptical Science*, skepticalscience.com/ice-age-predictions-in-1970s-basic.htm

11. Annual climate statement of the Australian Bureau of Meteorology (2019): bom.gov.au/climate/current/annual/aus/2019

12. Geert Jan van Oldenborgh, Folmer Krikken, Sophie Lewis, et al., "Attribution of the Australian bushfire risk to anthropogenic climate change," *Natural Hazards and Earth Science Systems* 21, no. 3 (2021): 941–960, doi.org/10.5194/nhess-21-941-2021

13. Stephen J. Pyne, "The Pyrocene comes to Australia: A commentary," *Journal & Proceedings of the Royal Society of New South Wales* 153, no. 1 (2020): 30–35, royalsoc.org.au/images/pdf/journal/153-1-Pyne.pdf

14. Geert Jan van Oldenborgh, Folmer Krikken, Sophie Lewis, et al., "Attribution of the Australian bushfire risk to anthropogenic climate change," *Natural*

Hazards and Earth Science Systems 21, no. 3 (2021): 941–960, doi.org/10.5194/nhess-21-941-2021

15. Gabi Mocatta and Erin Hawley, "Uncovering a climate catastrophe? Media coverage of Australia's Black Summer bushfires and the revelatory extent of the climate blame frame," *M/C Journal* 23, no. 4 (2020), doi.org/10.5204/mcj.1666

16. Ibid.

17. Ibid.

18. Ibid.

19. Ibid.

20. Ibid.

21. Ibid.

22. Ibid.

23. See tweets from @RoyPentland (x.com/RoyPentland/status/1223873572872916994?s=20) and @PRGuy17 (x.com/PRGuy17/status/1308012297952964608?s=20)

24. Christine Meyer, "Jörg Kachelmann äußert sich zu Waldbränden in Kanada," *Kölner Stadt-Anzeiger*, June 14, 2023, ksta.de/panorama/joerg-kachelmann-aeussert-these-zu-waldbraenden-in-kanada-1-590495

25. Adrian Brügger, Christina Demski, and Stuart Capstick, "How personal experience affects perception of and decisions related to climate change: A psychological view," *Weather, Climate, and Society* 13, no. 3 (2021): 397–408, doi.org/10.1175/WCAS-D-20-0100.1

26. Climate Change Litigation Databases, "Bushfire Survivors for Climate Action Incorporated v. Environmental Protection Authority" (2020), climate-laws.org/geographies/australia/litigation_cases/bushfire-survivors-for-climate-action-incorporated-v-environment-protection-authority

27. Ibid.

28. Ibid.

29. Rupert F. Stuart-Smith, Friederike E. L. Otto, Aisha I. Saad, et al., "Filling the evidentiary gap in climate litigation," *Nature Climate Change* 11 (2021): 651–655, doi.org/10.1038/s41558-021-01086-7

30. See: climatecasechart.com/wp-content/uploads/non-us-case-documents/2020/20201104_2021-NSWLEC-92-2020-NSWLEC-152-leave-granted-to-BSCA-to-file-and-serve-expert-evidence-on-climate-change-science_decision-1.pdf

31. Climate Change Litigation Databases, "Gloucester Resources Limited v. Minister for Planning" (2017), climatecasechart.com/non-us-case/gloucester-resources-limited-v-minister-for-planning

32. Green Legal Impact: www.greenlegal.eu/en/start-english

33. Benjamin H. Strauss, Philip M. Orton, Klaus Bittermann, et al., "Economic damages from Hurricane Sandy attributable to sea level rise caused by anthropogenic climate change," *Nature Communications* 12: 2720 (2021), doi.org/10.1038/s41467-021-22838-1

34. "Fairchild v Glenhaven Funeral Services," *LawTeacher* (2021), lawteacher.net/cases/fairchild-v-glenhaven-funeral-services.php

35. Sora Park, Caroline Fisher, Jee Young Lee, et al., "Digital news report: Australia 2020," News & Media Research Centre, University of Canberra, apo.org.au/node/305057

36. Ibid.

CHAPTER 8: GUILT AND RESPONSIBILITY

1. Christian Parth and Lisa Caspari, "Die Ahnungslosen," *Zeit Online*, July 13, 2022, zeit.de/politik/deutschland/2022-07/flut-untersuchungsausschuss-hochwasserkatastrophe-nrw-rlp

2. Sofia Martyanowa, "Entsetzt und verzweifelt: Augenzeugen berichten von Hochwasser-Katastrophe," *SNA*, July 20, 2021, web.archive.org/web/20211207040532/https:/snanews.de/20210720/entsetzt-verzweifelt-augenzeugen-berichten-von-hochwasser-katastrophe-2911888.html

3. Ira Schaible, "Ein Jahr nach der Flutkatastrophe: 'Die Aufbaueuphorie ist verflogen'," *GEO*, July 11, 2022, geo.de/wissen/ahrtal--1-jahr-nach-der-flutkatastrophe-32529436.html

4. Christina Nover, "Die Wut nach der Flut: 'Tod von Johanna war unnötig,'" *SWR Aktuell*, July 13, 2022, swr.de/swraktuell/rheinland-pfalz/koblenz/johanna-orth-bad-neuenahr-flutopfer-100.html

5. Alessa J. Truedinger, Ali Jamshed, Holger Sauter, and Joern Birkmann, "Adaptation after extreme flooding events: Moving or staying? The case of the Ahr Valley in Germany," *Sustainability* 15, no. 2 (2023), doi.org/10.3390/su15021407

6. Jordis S. Tradowsky, Sjoukje Y. Philip, Frank Kreienkamp, et al., "Attribution of the heavy rainfall events leading to severe flooding in Western Europe during July 2021," *Climatic Change* 176 (2023), doi.org/10.1007/s10584-023-03502-7

7. EM-DAT, the International Disaster Database: emdat.be

8. "Gesetz zur Errichtung eines Sondervermögens 'Aufbauhilfe 2021,'" September 10, 2021, bundesregierung.de/resource/blob/974430/1960344/14187d88e595078e560929ab674465f5/2021-09-15-gesetz-aufbauhilfe-data.pdf

9. Michael Dietze, Rainer Bell, Ugur Ozturk, et al., "More than heavy rain turning into fast-flowing water—a landscape perspective on the 2021 Eifel floods," *Natural Hazards and Earth System Sciences* 22, no. 6 (2022), doi.org/10.5194/nhess-22-1845-2022

10. Press release from the Koblenz district attorney's office (July 7, 2023): aktiplan.de/staatsanwaltschaft-zum-sachstand-im-ermittlungsverfahren-ahrflut

11. Annegret H. Thieken, Philip Bubeck, Anna Heidenreich, et al., "Performance of the flood warning system in Germany in July 2021—insights from affected residents," *Natural Hazards and Earth System Sciences* 23, no. 2 (2023), doi.org/10.5194/nhess-23-973-2023

12. Ibid.

13. Erin Coughlan de Perez, Kristoffer B. Berse, Lianne Angelico C. Depante, et al., "Learning from the past in moving to the future: Invest in communication and response to weather early warnings to reduce death and damage," *Climate Risk Management* 38 (2022), doi.org/10.1016/j.crm.2022.100461

14. Ibid.

15. Greta Thunberg, "We now have to do the seemingly impossible," *The Climate Book* (Penguin, 2023): 359, theclimatebook.org

16. Pauline Büsken, "Wie reagieren die Fraktionen im Bundestag auf die Fridays for Future Bewegung?" (2019), regierungsforschung.de/wie-reagieren-die-fraktionen-im-bundestag-auf-die-fridays-for-future-bewegung

17. Werner Eckert, "'Das Klimabuch' von Greta Thunberg: Weise und infantil zugleich," *SWR2*, October 26, 2022, web.archive.org/web/20221028225807/https://www.swr.de/swr2/literatur/das-klimabuch-von-greta-thunberg-weise-und-infantil-zugleich-100.html

18. Kate Connolly and Matthew Taylor, "Extinction Rebellion founder's Holocaust remarks spark fury," *The Guardian*, November 20, 2019, theguardian.com/environment/2019/nov/20/extinction-rebellion-founders-holocaust-remarks-spark-fury

    "Extinction Rebellion: Co-founder apologises for Holocaust remarks," *BBC News*, November 21, 2019, bbc.co.uk/news/world-europe-50501941

19. Christine Eriksen, Andrin Hauri, and David Nicolai Kollmann, "Adapting civil protection to a changing climate," *Policy Perspectives* 10, no. 12 (2022), ethz.ch/content/dam/ethz/special-interest/gess/cis/center-for-securities-studies/pdfs/PP10-12_2022-EN.pdf

20. Max Liboiron, Manuel Tironi, and Nerea Calvillo, "Toxic politics: Acting in a permanently polluted world," *Social Studies of Science* 48, no. 3 (2018): 331–349, doi.org/10.1177/0306312718783087

CHAPTER 9: A COUNTRY DROWNING IN CLIMATE DAMAGE

1. NPR interview with Steve Inskeep and Dr. Imran Lodhi: "A doctor in Punjab province describes relief efforts for Pakistan's floods," August 31, 2022, npr. org/2022/08/31/1120223451/a-doctor-in-punjab-province-describes-relief-efforts-for-pakistans-floods

2. Saima Mohsin, "Pakistan floods leave people desperate and hungry with warning more is to come," *Sky News* (2022), news.sky.com/story/pakistan-floods-leave-people-desperate-and-hungry-with-warning-more-is-to-come-12689655#

3. Hannah Ellis-Petersen and Shah Meer Baloch, "Pakistani PM says he should not have to beg for help after catastrophic floods," *The Guardian*, October 6, 2022, theguardian.com/world/2022/oct/06/pakistani-pm-says-he-should-not-have-to-beg-for-help-after-catastrophic-floods

4. Pakistan's seventh population and housing census (2023): pbs.gov.pk/sites/default/files/population/2023/Press%20Release.pdf

5. Friederike E. L. Otto, Mariam Zachariah, Fahed Saeed, et al., "Climate change increased extreme monsoon rainfall, flooding highly vulnerable communities in Pakistan," *Environmental Research: Climate* 2, no. 2 (2023), doi.org/10.1088/2752-5295/acbfd5

6. Ibid.

7. OCHA Humanitarian Advisory Team, "Pakistan: 2022 monsoon floods—situation report no. 5," September 9, 2022, reliefweb.int/report/pakistan/pakistan-2022-monsoon-floods-situation-report-no-5-9-september-2022

8. Germanwatch, "Global Climate Risk Index 2021," germanwatch.org/en/cri

9. Jeff Ridley, Andrew Wiltshire, and Camilla Mathison, "More frequent occurrence of westerly disturbances in Karakoram up to 2100," *Science of the Total Environment* 468–469 (2013): S31–S35, doi.org/10.1016/j.scitotenv.2013.03.074

10. H. Douville, K. Raghavan, J. Renwick, et al., "Water cycle changes," *Climate Change 2021: The Physical Science Basis* (2023): 1055–1210, doi.org/10.1017/9781009157896.010

11. "The flood 2010," report by the Flood Inquiry Commission of Pakistan, pakissan.com/english/watercrisis/flood/report.of.flood.inquiry.commission.shtml

12. Notre Dame Global Adaptation Index (ND-GAIN): gain.nd.edu/our-work/country-index/methodology

13. Asad Sarwar Qureshi, "Water management in the Indus Basin in Pakistan: Challenges and opportunities," *Mountain Research and Development* 31, no. 3 (2011): 252–260, doi.org/10.1659/MRD-JOURNAL-D-11-00019.1

14. Pakistan Bureau of Statistics website: pbs.gov.pk

15. Dilshad Ahmad and Muhammad Afzal, "Climate change adaptation impact on cash crop productivity and income in Punjab province of Pakistan," *Environmental Science and Pollution Research* 27 (2020): 30767–30777, doi.org/10.1007/s11356-020-09368-x

16. Friederike E. L. Otto and Frederick Fabian, "Equalising the evidence base for adaptation and loss and damages," *Global Policy* (2023), doi.org/10.1111/1758-5899.13269

17. See the Paris Agreement: 9–11: https://unfccc.int/files/essential_background/convention/application/pdf/english_paris_agreement.pdf.
Also see the United Nations Framework Convention on Climate Change (1992).

18. Federal Ministry for the Environment, Nature Conservation, Nuclear Safety and Consumer Protection, "German strategy for adaptation to climate change summary," bmuv.de/en/download/german-strategy-for-adaptation-to-climate-change-summary

19. See Copenhagen Accord: unfccc.int/files/meetings/cop_15/application/pdf/cop15_cph_auv.pdf

20. See Paris Agreement: 12–13: unfccc.int/files/essential_background/convention/application/pdf/english_paris_agreement.pdf

21. Emily Boyd, Rachel A. James, Richard G. Jones, et al., "A typology of loss and damage perspectives," *Nature Climate Change* 7 (2017): 723–729, nature.com/articles/nclimate3389

22. Simon Fraser, "Pakistan floods are 'a monsoon on steroids,' warns UN chief," *BBC News*, August 30, 2022, bbc.co.uk/news/world-asia-62722117. Georgina Rannard, "How Pakistan floods are linked to climate change," *BBC News,* September 2, 2022, bbc.co.uk/news/science-environment-62758811. Damian Carrington, "'Monster monsoon': Why the floods in Pakistan are so devastating," *The Guardian*, August 29, 2022, theguardian.com/environment/2022/aug/29/monster-monsoon-why-the-floods-in-pakistan-are-so-devastating

23. Friederike E. L. Otto, Mariam Zachariah, Fahad Saeed, et al., "Climate change increased extreme monsoon rainfall, flooding highly vulnerable communities in Pakistan," *Environmental Research: Climate* 2, no. 2 (2023), doi.org/10.1088/2752-5295/acbfd5

24. Chao Li, Francis Zwiers, Xuebin Zhang, et al., "Changes in annual extremes of daily temperature and precipitation in CMIP6 models," *Journal of Climate* 34, no. 9 (2021): 3441–3460, doi.org/10.1175/JCLI-D-19-1013.1

25. See Article 8, Paragraph 52 of the Paris Agreement: unfccc.int/files/adaptation/groups_committees/loss_and_damage_executive_committee/application/pdf/ref_8_decision_xcp.21.pdf

26. "Secretary-General's address to the General Assembly," September 20, 2022, un.org/sg/en/content/sg/statement/2022-09-20/secretary-generals-address-the-general-assembly-trilingual-delivered-follows-scroll-further-down-for-all-english-and-all-french

27. "Bramble Cay melomys: Climate change-ravaged rodent listed as extinct," *BBC News*, February 20, 2019, bbc.com/news/world-australia-47300992.amp

28. ClientEarth, "Torres Strait Islanders win historic human rights legal fight against Australia," September 23, 2022, clientearth.org/latest/press-office/press/torres-strait-islanders-win-historic-human-rights-legal-fight-against-australia

29. P. Tschakert, N. R. Ellis, C. Anderson, et al., "One thousand ways to experience loss: A systematic analysis of climate-related intangible harm from around the world," *Global Environmental Change* 55 (2019): 58–72, doi.org/10.1016/j.gloenvcha.2018.11.006

30. Report by Working Group I, IPCC Sixth Assessment Report, doi.org/10.1017/9781009157896

31. Ayesha Tandon, "Analysis: How the diversity of IPCC authors has changed over three decades," *Carbon Brief*, March 15, 2023, carbonbrief.org/analysis-how-the-diversity-of-ipcc-authors-has-changed-over-three-decades

32. John Bohannon, "How much did your university pay for your journals?," *Science*, June 16, 2014, science.org/content/article/how-much-did-your-university-pay-your-journals

33. Christopher Ketcham, "How scientists from the 'Global South' are sidelined at the IPCC," *The Intercept*, November 17, 2022, theintercept.com/2022/11/17/climate-un-ipcc-inequality

34. Jason Hickel and Aljosa Slamersak, "Existing climate mitigation scenarios perpetuate colonial inequalities," *The Lancet Planetary Health* 6, no. 7 (2020): e628–e631, doi.org/10.1016/S2542-5196(22)00092-4

    Also see Jason Hickel and Giorgos Kallis, "Is green growth possible?," *New Political Economy* 25, no. 4 (2020), doi.org/10.1080/13563467.2019.1598964

35. Ibid.

36. Description of the UNCC transitional committee: unfccc.int/topics/adaptation-and-resilience/groups-committees/transitional-committee

37. UNFCCC draft report: unfccc.int/sites/default/files/resource/CP2022_L0IE.pdf

38. Synthesis report on the technical dialogue of the first global stocktake of the UNFCCC, unfccc.int/documents/631600#

39. Harro van Asselt and Fergus Green, "COP26 and the dynamics of anti-fossil fuel norms," *WIREs Climate Change* 14, no. 3 (2023), doi.org/10.1002/wcc.816

40. Daisy Dunne, Josh Gabbatiss, Aruna Chandrasekhar, et al., "Q&A: Should developed nations pay for 'loss and damage' from climate change?," *Carbon Brief*, September 26, 2022, interactive.carbonbrief.org/q-a-should-developed-nations-pay-for-loss-and-damage-from-climate-change

41. Tahura Farbin and Saleemul Huq, "Designing a comprehensive institutional structure to address loss and damage from climate change in Bangladesh," March 1, 2021, lossanddamagecollaboration.org/stories/designing-a-comprehensive-institutional-structure-to-address-loss-and-damage-from-climate-change-in-bangladesh

42. African Risk Capacity. Update posted on August 12, 2022, arc.int/news/african-risk-capacity-launches-its-first-parametric-insurance-product-against-high-impact

CHAPTER 10: WHAT NOW?

1. "Limited data prevent assessment of role of climate change in deadly floods affecting highly vulnerable communities around Lake Kivu," World Weather Attribution, June 29, 2023, worldweatherattribution.org/limited-data-prevent-assessment-of-role-of-climate-change-in-deadly-floods-affecting-highly-vulnerable-communities-around-lake-kivu

2. Infratest dimap, "Sonntagsfrage Bundestagswahl," infratest-dimap.de/umfragen-analysen/bundesweit/sonntagsfrage

3. V. Pawlik, "Wie bewerten die Klimaproteste der Gruppe 'Letzte Generation,'" Statista, de.statista.com/statistik/daten/studie/1345160/umfrage/bewertung-der-letzte-generation-klimaproteste

4. Ellen Ehni, "Klimawandel als wichtigstes Problem," *tagesschau.de*, April 6, 2023, tagesschau.de/inland/deutschlandtrend/deutschlandtrend-3339.html

5. Peter Norton, "The hidden history of American anti-car protests," *Bloomberg*, October 8, 2019, bloomberg.com/news/articles/2019-10-08/the-hidden-history-of-american-anti-car-protests

6. Robert Costanza, Ida Kubiszewski, Enrico Giovannini, et al., "Development: Time to leave GDP behind," *Nature* 505 (2014): 283–285, doi.org/10.1038/505283a

7. "Reducing inequality benefits everyone—so why isn't it happening?," *Nature* 620 (2023): 468, doi.org/10.1038/d41586-023-02551-3

8. Lucy Cooke, *Bitch* (Transworld Publishers, 2023)

9. Rebecca Solnit, "What if climate change meant not doom—but abundance?," *The Washington Post*, March 15, 2023, washingtonpost.com/opinions/2023/03/15/rebecca-solnit-climate-change-wealth-abundance

Index

adaptation: challenges to, 32, 69–70; conditions for success, 72, 80, 102–4; extreme heat and, 32, 67–70, 77–79; fires and, 146–47; international climate policy and, 205–7; and loss and damage, 210, 221; Notre Dame Global Adaptation Initiative (ND-GAIN) index, 203–4; physiological, 68–69

Africa: attribution studies and, 213; climate models and, 217; development projects, 70–72; drought, 58–59, 91; extreme heat, 52–54, 55, 57–61, 65–66; fire, 60; media and, 64–66; migration, 131–32; and research and knowledge production, 63–64; weather-monitoring infrastructure and, 61–63. *See also specific countries*

African Risk Capacity Group, 222–23

agricultural (ecological) droughts, 91–92

Ahmedabad (India), 56–57, 68, 69

aid, development, 70–72, 123–26

air conditioning, 35–36

air pollution (air quality), 47–48, 60–61, 147, 164

air temperature readings, 62

Albanese, Anthony, 170

Alckmin, Geraldo, 102

Amazon, Brazilian: climate change and, 144, 145–46; deforestation, 139–40, 141–42, 147–48, 152, 154; fires (2019), 136, 137–39

Angelou, Maya, 233

Antarctic, 58

Argentina, 25–26, 144

art, 233

Asia, 91, 122. *See also specific countries*

attribution science (attribution studies): about, 4, 6–7; fires and, 145; German floods (2021), 184–85; Madagascar drought (2021), 119; Pacific Northwest heat wave (2021), 23, 28–30; Pakistan floods (2022), 211–13

Australia: bushfires (2019–20), 158, 162, 164–65, 168–69; climate activism, 169, 176–77; climate lawsuits, 169–73; coal and gas industry, 160–63; drought, 90, 91; Forest Fire Danger Index, 145; media, misinformation, and climate skepticism, 103, 158–59, 162–63, 165–67, 168

Bahamas, 208

Bangladesh, 67, 189, 222

Barroso, Luís Roberto, 152

Berejiklian, Gladys, 165

body, human, 68–69

Bolsonaro, Jair, 136, 138, 140–41, 142, 143, 150–51, 154

Botswana, 58

Brazil: Amazon fires (2019), 136, 137–39; climate change influence on fires, 145–46; climate damage attributed to, 142–43; climate lawsuits, 150–53, 154–55; deforestation, 139–41, 141–42, 147–48, 152, 154. *See also* São Paulo

British Columbia, 30, 37–38, 40–41. *See also* Lytton; Pacific Northwest heat wave (2021)

Canada, 14, 45, 56, 167–68. *See also* Fire Weather Index; Pacific Northwest heat wave (2021)

Cape Town, 86–87, 89, 93–94, 95–101, 104, 105, 110–12

capitalism, 15, 43–44, 45–49. *See also* colonial-fossil narrative

causality, 129, 155, 175

Chad, 204

Chenery, Hollis B., 123

Chevron, 150

China, 102

cities, 26, 46

climate change: approach to, 16–17; contradictory views about, 226–27; costs, 107–8; crime analogy, 229; denial, skepticism, and misinformation, 158–59, 162–63, 165–68, 226; and dignity and human rights, 10, 11–13; economic cost-benefit assessment, 151–52, 216; habituation to, 43–44, 196; migration and, 129–32, 224–25; need for new narratives, 196–99, 200, 228–33; politics and, 95–101, 102–3; social perception and, 155–56, 173–75, 226–27; teachable

moments, 4–5, 8; tipping points, 42–43, 228–29; wake-up calls, 39, 42, 56, 101–2. *See also* adaptation; colonial-fossil narrative; disasters, climate; droughts; fires; flooding; heat, extreme; inequality and injustice; justice; lawsuits, climate; loss and damage; media; research and knowledge production; temperature

climate models, 29–30, 217

climate science (climate studies), 9, 174–76. *See also* research and knowledge production

coal industry, 160–62, 172–73, 205

colonial-fossil narrative: about, 15–16; Cape Town's inequality and, 101, 112; justice system and, 156; need for counter-narratives, 196–99, 200, 228–33; prevalence and persistence of, 13–15, 45–49, 195–96, 220, 227–28; and research and knowledge production, 63, 217–18. *See also* inequality and injustice

colonialism: continued influence of, 9; and development aid and NGOs, 123–26; in Madagascar, 119–21; at UN, 222. *See also* decolonization

communications, 189–90

conflict and war, 12, 128–29, 225–26

COP (Conference of the Parties): about, 11; COP24 (Katowice, 2018), 205; COP26 (Glasgow, 2021), 207; COP27 (Sharm el-Sheikh, 2022), 205, 219; COP28 (Dubai, 2023), 214, 219–20. *See also* loss and damage

cost-benefit assessment, economic, 151–52, 216

Covid-19 pandemic, 12, 15, 34, 44, 102, 105, 116–17, 125, 199, 230

Cox, Brian, 159

deaths: air pollution and, 47–48;
Brazil's responsibility, 140–41,
142–43; forest fires and, 147, 162;
heat and, 33–35, 41–42, 55, 57, 69
decolonization, 194. *See also*
colonialism
deforestation, 139–40, 141–42, 147–
48, 152, 154
Delhi (India), 26, 60–61
Democratic Republic of the Congo,
225–26
development aid, 70–72, 123–26
dignity, 10, 11. *See also* human rights;
inequality and injustice; justice
disasters, climate: difficulty imagining, 24; responsibility for, 120–21,
121–23, 190–92; risk factors,
5–6
disinformation. *See* misinformation
domination, 9–10
droughts: climate change and, 90–91,
94; El Niño-Southern Oscillation
(ENSO) and, 143–44; fires and,
143; food security and, 91–92;
heat waves and, 58–59; Kenya,
118; Lesotho, 108–10; Madagascar,
7, 114–18, 119; São Paulo, 101–2;
South Africa, 92–95; types
of, 91
Dust Bowl, 2

Ebi, Kristie, 42
ecological (agricultural) droughts,
91–92
El Niño-Southern Oscillation (ENSO),
143–44, 146
Emergency Events Database
(EM-DAT), 51–52, 53–54, 55,
61–62, 63, 64, 65, 186
Eswatini, 89, 108
Europe: droughts, 90–91; gender
inequality, 122; heat waves, 33, 39,

55–56, 182, 189; media and, 64.
See also specific countries
Extinction Rebellion, 197, 198
extractivist, 15, 16. *See also* colonial-
fossil narrative
extreme weather events. *See* climate
change; droughts; fires; flooding;
heat, extreme; weather, extreme
ExxonMobil, 44, 150, 175

FEWS NET, 115–16
Fiji, 208
fires: adaption measures, 146–47; air
pollution from, 147; Australian
bushfires (2019-20), 158, 162,
164–65, 168–69; Brazilian Amazon (2019), 136, 137–39; Central
Africa, 60; climate change and,
145–46; contributing factors,
144–45; Lytton (BC), 22–23, 28,
41; media misinformation and,
165–68; Pacific Northwest heat
wave (2021) and, 40–41; psychological repercussions, 168–69;
"Pyrocene" age, 165
Fire Weather Index (FWI), 145–46
flooding: glacial lake outburst floods
(GLOFs), 59–60; Huaraz, Peruvian Andes (1941), 2; Pakistan,
59–60, 202–3, 203–5, 211–13;
Rwanda and Democratic Republic of the Congo (2023), 225–26;
West Africa (2022), 5–6. *See also*
Germany, flooding (2021)
Florida, 74–75
food (in)security, 91–92, 104–6, 108–
9, 110, 117–18
Forest Fire Danger Index, 145
forest fires. *See* fires
fossil-fuel phaseout, 220. *See also* coal
industry; colonial-fossil narrative
France, 39, 42, 119, 192–93

The Gambia, 52, 56, 57, 67, 73–77, 192
gender equality, 230. *See also* patriar-
    chy; women
Germany: climate justice shortfalls,
    192–93; climate lawsuits, 148,
    208; colonial-fossil narrative in,
    195–96; daily water usage, 87;
    development projects in Global
    South, 71; disaster protection sys-
    tems, 199–200; legal system and
    scientific input, 174; media and
    climate misinformation, 167–68;
    perceived lack of natural disas-
    ters, 182; reunification, 232; risk
    awareness and communication
    issues, 38, 103, 188–90; structural
    racism and patriarchy in, 193–94
Germany, flooding (2021): about,
    182, 183–85; climate change
    influence, 185–86; development
    factors, 186–88; responsibility
    for, 190–92; risk awareness and
    communication issues, 38, 103,
    188–90
Ghana, 120, 192
glacial lake outburst floods (GLOFs),
    59–60
Global North: adaptation and, 67;
    climate change perceptions, 151,
    174, 227; Covid-19 pandemic and,
    15; development aid from, 70–72,
    123–26; domination narrative and,
    9–10; fossil-fuel narrative and, 14;
    fossil-fuel phaseout and, 220; and
    loss and damage, 208, 213–14,
    219; and research and knowledge
    production, 63–64, 217–18. *See
    also* colonial-fossil narrative; colo-
    nialism; *specific countries*
Global South: adaptation in, 67–68,
    193; Covid-19 pandemic and,
    15; development aid to, 70–72,

123–26; domination narrative and,
    9–10; fossil-fuel narrative and, 14;
    and NGOs and structural racism,
    126–27; and research and knowl-
    edge production, 63–64, 217–18;
    weather-monitoring infrastruc-
    ture and, 61–63. *See also* Africa;
    specific countries
global warming. *See* climate change;
    temperature
Gloucester Resources Limited, 172–73
greenhouse gas emissions, 14, 52, 81,
    165, 206, 228
gross domestic product (GDP), 227
groundwater (hydrological) droughts,
    91
Guterres, António, 214

habituation, 43–44, 196
Hallam, Roger, 198
Harari, Yuval Noah: *Sapiens*, 48
heat, extreme: adaptation to, 32,
    67–70, 77–79; Africa, 52–54, 55,
    57–61, 65–66; air quality and,
    60–61; cities and, 26, 46; climate
    change and, 2, 7, 24–26, 28–30,
    33, 53; costs, 61–62, 63–64;
    deaths from, 33–35, 41–42, 55,
    57, 69; development projects and,
    70–72; as deviation from mean,
    57–58; drought and, 58–59; early-
    warning systems, 79; Europe, 33,
    39, 55–56, 182, 189; fast pace
    of, 52, 79–81; female farmers in
    The Gambia and, 73–77; fires
    and, 40–41, 60; forecasting vs.
    warning, 55–56; government
    failures, 36–38; heat action plans,
    38, 39, 46, 56–57, 68, 70, 71; heat
    stress (heat sickness), 53, 74–75;
    media and, 40–41, 46–47, 64–66;
    meltwater floods and, 59–60;

migration and, 131–32; *The Ministry for the Future* (Robinson) and, 37; physics of, 57–61; and research and knowledge production, 63–64, 72–73, 78, 80–81; vulnerability to, 32–33; wake-up calls and tipping points, 39, 41–42, 56; weather-monitoring infrastructure and, 61–63. *See also* Pacific Northwest heat wave (2021)

Heggie, Jon, 86, 88–89

Hillman, Ralph, 160

Hispar Glacier, 59–60

housing, right to, 112

Huaraz (Peruvian Andes), 2

human rights, 10, 11–13, 111, 148–49, 151–53, 154–55, 222. *See also* inequality and injustice; justice

Hurricane Harvey (2017), 107, 108

Hurricane Katrina (2005), 35

Hurricane Sandy (2012), 175

hydrological (groundwater) droughts, 91

India: climate change approach, 45, 220–21; Delhi air quality, 60–61; drought adaptation, 193; food insecurity, 106; heat adaptation, 56–57, 67; heat waves, 25, 33, 52, 65, 66

inequality and injustice: air pollution and, 47–48; climate change and, 8–9, 12–13, 16, 44–45, 45–47, 128–29, 175, 224, 226; need to understand, 43, 132; Pacific Northwest heat wave (2021) and, 39–41; vs. responsibility, 190–91; South Africa, 110–12; vulnerability and, 122. *See also* colonial-fossil narrative; colonialism; justice; racism, structural

infrastructure and institutions, 48, 61–63, 80. *See also* development aid

insurance industry, 51, 64, 223

Intergovernmental Panel on Climate Change (IPCC), 24–25, 77, 91, 119, 206, 216–17

International Criminal Court (ICC), 153–55

International Disaster Database (EM-DAT), 51–52, 53–54, 55, 61–62, 63, 64, 65, 186

Italy, 91

justice: and loss and damage, 208–11, 213–14, 221–23; need for new narratives, 196–99, 200, 228–33; pioneers and laggards, 192–93; toxic, 198. *See also* inequality and injustice; lawsuits, climate

Just Stop Oil, 197, 226

Kachelmann, Jörg, 167

Karachi (Pakistan), 69

Kennedy, Robert, 227

Kenya, 57, 59, 118

Klein, Naomi, 197

knowledge production. *See* attribution science; research and knowledge production

Last Generation (Letzte Generation), 197, 226

lawsuits, climate: about, 148–50; Australia, 169–73; Brazil, 150–53, 154–55; Germany, 148, 208; in international courts, 153–55; social perception and, 155–56, 173–74

Lesotho, 58, 89, 94, 105, 108–10, 115

Levant, 90–91

Levy, Adam, 197

Liboiron, Max, 200
Limberg, Xanthea, 99
Lindner, Christian, 195
Lliuya, Saúl Luciano, 149
London (U.K.), 7, 26–27, 47–48, 57,
    60–61, 174, 212
loss and damage: about, 205–6, 207–
    8; action outside of global climate
    talks, 222–23; breakthrough at
    COP conferences, 219–21; chal-
    lenges quantifying, 213; exclusion
    from scientific studies, 216–17;
    justice and, 208–11, 213–14, 221–
    23; noneconomic loss and damage
    (NELD), 214, 215–16
Lula da Silva, Luiz Inácio, 139–40,
    142, 148, 153
Lytton (BC), 22–23, 28, 31, 41

Madagascar: climate change misat-
    tributed, 118–19, 122–23; climate
    change vulnerabilities, 127–28;
    colonialism and, 8, 119–21, 126;
    drought, poverty, and food inse-
    curity, 7, 114–18; migration and,
    130, 224
Mauritius, 208
McArthur, Edward, 40
McCormack, Michael, 165
media: Africa and, 64–66; climate
    change reporting, 229; and cli-
    mate denial, skepticism, and
    misinformation, 158–59, 162–63,
    165–68; extreme heat and, 40–41,
    46–47, 64–66; importance of
    independent media, 103; on num-
    bers, 30–31; on Pacific Northwest
    heat wave (2021), 23–24, 40–41;
    responsibility for climate crisis,
    191; tipping point discussions,
    228–29
Mediterranean, 90–91

meteorological droughts, 91
migration, 129–32, 224–25
misinformation, 103, 158–59, 162–63,
    165–68
mitigation, 205–6
Modi, Narendra, 220
Morrison, Scott, 167
Morrison, Toni, 233
mortality. See deaths
Morton, Douglas, 137
Mozambique, 58
Muller, Mike, 97–100
"Mums for Lungs" (activist group), 47
Munich Re, 63
Murdoch, Rupert, 159, 166, 167
Myanmar, 67

Nelson, Maggie, 197, 199
neoliberalism, 11, 15, 200. See also
    colonial-fossil narrative
Neubauer, Luisa, 148, 195
News Corp, 159, 166
New York, 61, 147, 175
Nigeria, 6, 8, 52, 65, 214, 216
noneconomic loss and damage
    (NELD), 214, 215–16
nongovernmental organizations
    (NGOs), 55, 62, 122–23, 126–27
Nordhaus, William, 107
Norway, 204
Notre Dame Global Adaptation Initia-
    tive (ND-GAIN) index, 203–4
numbers, 30–31

ocean temperatures, 144
Oladosu, Adenike, 67
once-in-a-century events, 27

Pacific Northwest heat wave (2021):
    adaptive measures and warnings,
    31–32, 35–38; attribution study,
    23, 28–30; heat-related deaths,

34–35; justice issues, 39–41; lesson from, 44; Lytton (BC), 22–23, 28, 31, 41; as tipping point, 42; as unprecedented, 23–24

Pakistan: climate change approach, 221; flooding, 59–60, 202–3, 203–5, 211–13; food insecurity, 106; heat adaptation, 67; heat waves, 25, 65–66, 69; and loss and damage, 223

Paris Agreement (2015), 3, 12, 139, 151–53, 205–8, 210, 213, 218

patriarchy, 8–9, 72, 193–94, 217

phaseout, fossil-fuel, 220

Piketty, Thomas, 197

politics, 95–101, 102–3

pollution, air, 47–48, 60–61, 147, 164

Portland, 36. See also Pacific Northwest heat wave (2021)

power structures, 126, 199, 200

Preston, Brian, 170–73

"Pyrocene" age, 165

Quebec, 61, 147

racism, structural, 126–27, 192, 193–94

Rehman, Sherry, 211

research and knowledge production: bias against and exclusion of Global South, 63–64, 217–18; climate models, 29–30, 217; credibility issues, 9; extreme heat and, 72–73, 78, 80–81; and loss and damage, 216–17; migration and, 129–30; need to disseminate climate research, 174–76; teachable moments and, 4–5; uncertainty range, 31. See also attribution science

resilience, 79–80, 128

responsibility, for disasters, 120–21, 121–23, 190–92

Richard Toll (Senegal), 124

risk and risk awareness, 5–6, 188–89

Robinson, Kim Stanley: The Ministry for the Future, 37

Russia, 33, 66, 125, 227. See also Siberia

Rwanda, 225–26

RWE, 44, 149–50, 222

Sackett, Penny, 169–70

Sahel region, 124, 128–29

São Paulo, 96, 101–2, 137

Saudi Arabia, 52

Saudi Aramco, 48–49

scientific studies. See research and knowledge production

Scotland, 57

Seattle, 36, 40. See also Pacific Northwest heat wave (2021)

Senegal, 124

Setzer, Alberto, 137–38, 138–39

Shklar, Judith, 155

shock doctrine, 197

Siberia, 7, 33, 216

Silva, Marina, 139–40

Singh, Harjeet, 221

social perception, 155–56, 173–75, 226–27. See also tipping points

Solnit, Rebecca, 231

Somalia, 118

South Africa: Cape Town water crisis, 86–87, 89, 93–94, 95–101, 104, 105, 110–12; climate change and politics, 95–101; drought vulnerability, 92–95; extreme heat, 52, 58; housing rights, 112; water precarity, 87–89; water rights and inequality, 111–12

South Asia, 25, 59, 106, 212. See also Bangladesh; India; Pakistan

Sparks, Carol, 165–66

Statista, 63

Sudan, 52
Switzerland, 149

Taylor, Angus, 166–67
teachable moments, 4–5, 8
temperature: 1.5°C (2.7°F) warmer, 3, 11, 32, 53, 141, 195, 220; 2°C (3.6°F) warmer, 11, 32, 94, 151; acceptable limits debate, 2–4; air temperature readings, 62; global average, 1–2, 24–25; ocean temperature, 144. *See also* heat, extreme
Thunberg, Greta, 191, 195; *The Climate Book*, 195–96
time, 104
tipping points, 42–43, 228–29
tropical storms, 64
Trudeau, Justin, 14
Trump, Donald, 162

Uganda, 120
Ukraine, 66, 105, 125, 202, 227
uncertainty range, 31
United Kingdom, 35, 45, 87, 142, 182, 212. *See also* London
United Nations, 148, 222. *See also* COP (Conference of the Parties)
United Nations Department of Economic and Social Affairs (UN DESA), 77
United Nations Framework Convention on Climate Change (UNFCCC), 209, 210
United Nations Office for Disaster Risk Reduction (UNDRR), 5
United States of America, 45, 64, 147, 150, 162, 220. *See also* Pacific Northwest heat wave (2021)
Unlearn Patriarchy (Ryland, Jaspers, and Horch), 194
urban environments (cities), 26, 46

Vancouver, 35, 36, 39. *See also* Pacific Northwest heat wave (2021)
vulnerability: to climate change in Madagascar, 127–28; to drought in South Africa, 92–95; European blindness to, 192; to extreme heat, 32–33; extreme weather risk and, 5–6; to flooding in Germany, 188–89; to flooding in Pakistan, 203–5; migration and, 130; social and political factors, 122

wake-up calls, 39, 42, 56, 101–2
Wallace-Wells, David, 23–24
war and conflict, 12, 128–29, 225–26
Warsaw International Mechanism, 207
water: Cape Town crisis, 86–87, 89, 93–94, 95–101, 104, 105, 110–12; daily usage comparisons, 86–87; right to, 111–12; South African precarity, 87–89. *See also* droughts; flooding
weather, extreme: forecasting vs. warning, 55–56; statistical modelling, 26–28. *See also* climate change; droughts; fires; flooding; heat, extreme
Weekes-Richemond, Lazenya, 126–27, 198
women: female farmers in The Gambia, 73–77; gender equality, 230; inequality against, 122; leadership in climate change debates, 78; white women in NGOs, 126–27
World Bank, 124, 125, 202
World Food Programme (WFP), 114, 118–19

Zimbabwe, 58

**DAVID
SUZUKI
INSTITUTE**

THE DAVID SUZUKI INSTITUTE is a companion organization to the David Suzuki Foundation, with a focus on promoting and publishing on important environmental issues in partnership with Greystone Books.

We invite you to support the activities of the Institute. For more information, please write to us at info@davidsuzukiinstitute.org.